Sustainable Quality

Sustainable Quality presents the implementation and operation of systematic quality management from a project management perspective, demonstrating how project management concepts can strengthen the development, governance, and continual improvement of management systems.

In this Second Edition, the authors explore the origins of quality management as a discipline, its appearance in its present form, and how systematic quality management can be implemented and applied in all kinds of organisations to achieve stability and better results. The basic principles of quality management and the ISO 9001 quality management standard are discussed and explained from a broad perspective, with illustrative examples from different types of organisations.

The book offers a global, accessible guide for undergraduate and postgraduate university students. Written clearly and with illustrative examples, it will also appeal to all those interested in project management and quality management and wishing to expand their knowledge base.

Helgi Thor Ingason is Professor at Reykjavik University, Iceland, where he chairs the Master of Project Management (MPM) programme and the Sustainability Institute & Forum (SIF). He is an IPMA Certified Executive Trainer in Project Management, an IPMA Certified Senior Project Manager, and a management consultant. He holds a PhD from the Norwegian University of Science and Technology (NTNU), an MSc from the University of Iceland, and a Stanford Advanced Project Management certification. His research focuses on project management, quality management, sustainability, and energy systems. Helgi Thor is the author and co-author of several books in English and Icelandic and has published his research in several international journals. He has also served in several senior leadership positions, including Research Coordinator of IPMA and interim CEO of Reykjavik Energy and SORPA.

Jaakko Kujala is Professor of Project and Quality Management at the Industrial Engineering and Management Research Unit, University of Oulu, Finland. He holds an MSc in Information Technology and a Doctor of Technology degree in Industrial Engineering and Management from the Helsinki University of Technology. He has over ten years of practical experience in developing quality management systems in different roles in industry. His research and teaching have focused on how different management approaches and methods should be applied in ways that take into account the specific context, needs, and characteristics of organisations. His current research interests also include how organisations can make effective and responsible use of artificial intelligence in their operations, management practices, and organisational development. He has authored more than 150 scientific publications in international journals, conferences, and books.

Sustainable Quality
A Project Management Perspective
Second Edition

Helgi Thor Ingason and Jaakko Kujala

Routledge
Taylor & Francis Group
LONDON AND NEW YORK

Designed cover image: Árni Sigurður Ingason

Second edition published 2027
by Routledge
4 Park Square, Milton Park, Abingdon, Oxon, OX14 4RN

and by Routledge
605 Third Avenue, New York, NY 10158

Routledge is an imprint of the Taylor & Francis Group, an informa business

First edition published by Routledge 2020

British Library Cataloguing-in-Publication Data
A catalogue record for this book is available from the British Library

ISBN: 9781041138341 (hbk)
ISBN: 9781041138334 (pbk)
ISBN: 9781003671749 (ebk)

DOI: 10.4324/9781003671749

Typeset in Sabon
by Deanta Global Publishing Services, Chennai, India

Contents

Acknowledgements

This book builds on many years of collaboration with students, practitioners, colleagues, and organisations working with quality management, project management, and systematic organisational development. We are deeply grateful for the many discussions, projects, challenges, and reflections that have shaped our thinking over the years.

We would first like to acknowledge the late Professor Pétur K. Maack, whose teaching and pioneering work in quality management inspired an early interest in the field and helped lay an important foundation for this book. We would also like to acknowledge Professor Paul Lillrank, whose thinking has contributed to our understanding of how quality management can be applied thoughtfully across different types of organisations and processes. We are also thankful to all those who contributed valuable feedback, comments, and suggestions during the development of the manuscript.

We would particularly like to thank Dr. Jóhanna Gunnlaugsdóttir, Dr. Þórður Víkingur Friðgeirsson, Dr. Haukur Ingi Jónasson, Guðmundur S. Pétursson quality manager, Pétur Helgason ISO auditor, Sveinn V. Ólafsson MSc mechanical engineer, and Agnes Hólm Gunnarsdóttir MSc industrial engineer for their thoughtful comments and constructive feedback. We are also grateful to Jonathan Norman for excellent English proofreading and valuable reflections on the manuscript.

Special thanks are extended to our students. The discussions with students in the Master of Project Management (MPM) programme at Reykjavik University, as well as students in quality management courses within the engineering department, have continuously challenged our thinking and provided important practical and academic perspectives.

We would also like to acknowledge Halldór Baldursson for the original illustrations developed for earlier versions of this book. New illustrations in this edition were developed with support from generative AI tools, primarily through iterative dialogue, refinement, and conceptual development between the authors and the AI interface.

Generative AI tools, particularly ChatGPT, were also used during the development of the manuscript as support for discussions, idea development, language refinement, structuring, and illustration design. The process was highly iterative and always guided by the authors' reflections, judgement, and experience. Working with these tools also provided valuable insights into the emerging role of generative AI in education and professional practice.

Finally, we would like to thank our families for their patience, support, and encouragement during the writing process. Helgi would also like to acknowledge his late father, Ingi Viðar Árnason, who read earlier versions of the manuscript and provided thoughtful comments and encouragement over the years.

We hope that this book will support students, professionals, and organisations in their continuous efforts to improve, learn, adapt, and contribute responsibly in a rapidly changing world.

Reykjavik and Oulu, 2026
Helgi Thor Ingason
Jaakko Kujala

Introduction
Why are we writing this book?

It is a fair question. Bookshelves are already full of books that explain quality principles, standards, tools, and systems. Many of them are excellent. Therefore, why did we feel the need to write this one? To answer that question, we need to explain where we come from.

We both work in academia, but our backgrounds are rooted in engineering and shaped by years of engagement with industry, operations, organisational development, project studies, teaching, and the practical realities of managing change in different settings. We share a passion not only for understanding organisations but also for helping them improve. We are deeply interested in developing new ideas and understanding how ideas become action, how strategies become practice, and how organisations can develop the capability to succeed in changing environments. This naturally connects quality management with project management, because modern organisations increasingly operate in projectified societies where change is organised through projects.

Our professional paths have been different, but they have also been complementary. Both of us have strong roots in industry and engineering, but we entered the field of quality management from somewhat different perspectives. One perspective emerged from highly technical industrial process environments and later evolved towards project management, organisational development, and systematic management. The other grew from information technology, project-oriented organisations, and the study of quality in complex industrial contexts. This difference has been a strength. It has allowed us to challenge assumptions, combine perspectives, and approach quality management as something deeply connected to projects, systems, people, technology, and organisational change.

The first version of this book was rooted in earlier Icelandic work on quality management. That foundation was valuable, but it also became clear that a broader international context required something more comprehensive. Organisations today operate in a world shaped by rapid technological development, increasing interconnectedness, growing sustainability pressures, geopolitical uncertainty, and rising stakeholder expectations regarding environmental and social responsibility. At the same time, the rapid emergence of generative AI is beginning to influence organisations in different ways. These realities demand critical reflection and new perspectives. This led to a collaboration aimed not simply at updating an existing book, but at rethinking what quality management education should offer today.

For Helgi Thor, this meant recognising that quality management must be more clearly connected to broader developments that shape management practice, the projectification of societies, the importance of a systems view, and an enhanced focus on sustainability and sustainable development. For Jaakko, it involved bringing fresh perspectives, critical reflections, and new dimensions related to learning, including the growing role of generative AI as a tool for teaching and capability development. Together, this collaboration became an opportunity to create something more relevant to the realities faced by modern organisations.

What has always been true is that organisations need to be financially sustainable and fulfil their broader roles in society. They must understand their customers, stakeholders, and expectations placed upon them. While these fundamental principles remain, the complexity has changed. The environment surrounding organisations is evolving rapidly, becoming more interconnected, and placing greater demands on leaders. Expectations extend far beyond traditional products or service quality. Health and safety, environmental responsibility, sustainability, resilience, transparency, and ethical legitimacy have become central issues. Organisations must not only satisfy customers, but also navigate broader stakeholder demands in an increasingly dynamic world. This is why we believe that quality management must be understood as part of something larger; systematic management.

In our view, organisations should not become servants of a management ideology, blindly applying tools, methods or standards without reflection. Instead, they must think critically, understand their own context, identify what matters most, and develop management systems that genuinely support their success. This requires not only an understanding of management principles and methods but also recognition of when and why specific practices are appropriate. Management practices must fit the context.

This belief shapes the entire philosophy of this book. We strongly believe that organisations benefit from systematic approaches that integrate quality, health, safety, and environmental management. Over time, this can evolve into an Integrated Management System (IMS), where these different dimensions work together in a coherent manner. Such integration supports sustainable management by aligning the organisational purpose, stakeholder expectations, operational control, and long-term responsibility.

This book is, therefore, about understanding quality management as a foundational discipline within the broader systematic and sustainable management. The structure of the book reflects this logic.

In Chapter 1, we introduce the fundamental concepts of quality management, as these concepts provide the foundation for everything that follows.

In Chapter 2, we explore the roots of the field, examining key pioneers, historical developments, and how quality thinking has evolved towards sustainability and integrated management.

Chapter 3 introduces systems thinking and systematic management, explaining how organisations can understand their purposes, stakeholder expectations, and risks through management systems.

Chapter 4 provides an overview of international standards and frameworks that help organisations structure their systematic efforts.

In Chapter 5, we emphasise implementation through projects, showing how management systems can be developed using a project approach across quality, health and safety, environmental management, or broader integrated contexts.

Chapter 6 focuses on the process approach, a cornerstone of quality management, and explains how processes can be understood, modelled, mapped, and improved.

Chapter 7 focuses on people, culture, communication, and information—the human and organisational dimensions that determine whether systems truly function in practice.

Chapter 8 addresses control, measurement, variation, auditing, and management review, because without understanding performance, control is impossible.

Finally, Chapter 9 focuses on improvement: the continuous quest to become better through structured problem-solving, learning, and capability development.

This final point may be the main reason for writing this book. Quality management, at its core, is about the relentless pursuit of improvement. It is about recognising that no system, organisation—or book—is ever truly finished. There is always room for learning, adaptation, refinement, and improvement. In this sense, writing this book has been an act of quality management.

Our ambition is to offer a practical, reflective, and context-sensitive guide that helps students, professionals, and leaders understand how systematic management can help organisations survive, improve, and contribute responsibly in a rapidly changing environment. The project perspective—already reflected in the title—signals our belief that organisations increasingly operate in a projectified world, where change, innovation, transformation, and strategic initiatives are frequently implemented through projects. A project approach is therefore not a secondary theme in this book, but an essential capability for organisations seeking to cope with rapidly changing environments.

If this book helps readers think more critically, act more systematically, and improve more sustainably, then it has fulfilled its purpose.

1 What is quality and why does it matter?

1.1 The concept of quality

We would like to begin with a personal example from Helgi Thor.

Figure 1.1 shows a photograph taken on a beautiful day in Iceland. It captures the landscape, the mountains, sky, lava, and moss. The photograph was taken by my brother, and it reminds me of how differently people can interpret the same object. Some may think of the purity of the air or the quality of the natural environment. Others may focus on the beauty of the landscape or even see spiritual meaning in the scene. A photographer might comment on technical aspects such as image resolution, exposure, aperture, focus, or perspective. What, then, does the concept of quality mean in this context?

It is tempting to describe the photograph as being of "high quality" or "low quality," or even to speak of the quality of the environment it shows. However, such statements

Figure 1.1 Photograph from Þórsmörk—a preserved area in south Iceland (Árni Sigurður Ingason).

DOI: 10.4324/9781003671749-1

immediately raise questions. High quality according to whom? Based on what criteria? Is there a universal scale for quality that everyone agrees on? Of course not. This simple example illustrates an important point; the word quality is vague when used in isolation. It becomes more meaningful when it is linked to expectations and requirements.

The term quality originates from the Latin word qualitas, which refers to character or nature. A widely cited and practical definition is that quality means *fitness for use*, a formulation introduced by Joseph M. Juran (1989).[1] This definition immediately shifts attention away from abstract judgements and towards a concrete question: fit for whose use, and under what expectations?

Juran further proposed that quality can be understood through two complementary aspects. The first is related to the features or characteristics of a product or service and what it is designed to deliver. For physical products, this may include features such as performance, reliability, durability, or functionality. For services, it may involve accuracy, competence, responsiveness, or availability. The features of a product or service have a significant impact on sales income. The more features that are built into it, the more likely customers are to accept a higher price. The second aspect concerns freedom from deficiency. This means delivering the product or service as intended, without errors, defects, or failures.

Customers form expectations regarding both. They expect a product or service to have certain characteristics and to be delivered consistently and without problems. If either aspect is missing, the product or service will not be perceived as fit for use, regardless of how impressive it may appear in other respects.

It is important to stress that quality is not about "more" or "less" in any absolute sense. This is about appropriateness. Quality exists only in relation to expectations and fitness for use. Therefore, a product or service can be said to have the right or wrong quality from the perspective of a particular user. This distinction is central to the book.

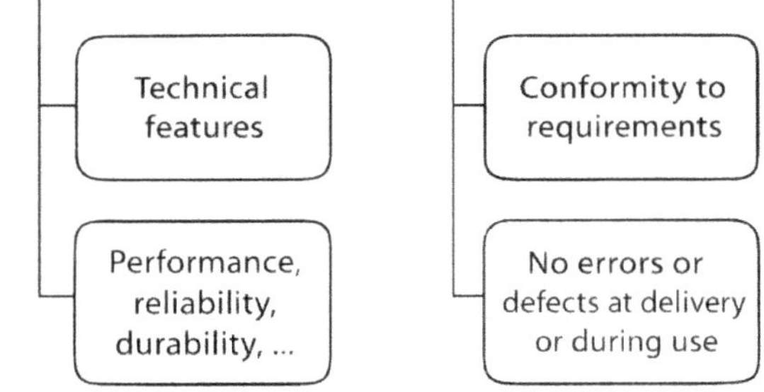

Figure 1.2 Two aspects of quality.

1.2 Why quality matters

Quality is perceived by some as a technical concern, delegated to specialists or confined to operational processes. From this perspective, quality management appears as an add-on—important, but secondary to e.g. strategy, leadership, and financial performance. This perspective is misleading.

Quality is a foundational management concept because it underpins the ability of an organisation to function, perform and be trusted. When products or services consistently meet expectations, customers develop confidence. When organisations deliver what they promise, they gain legitimacy. Over time, trust becomes a critical asset that enables stable relationships with customers, partners, regulators, and society at large. As a simple reflection, think about which grocery store you would choose when preparing for an important celebration—where having the right ingredients, of the right quality, available at the right time is essential for the dinner to be perfect.

Similarly, quality failures rarely remain isolated from one another. Defects, errors, or service breakdowns may appear as isolated problems with only immediate effects; however, they have underlying causes and consequences that often extend much further. Repeated failures can erode reputation, undermine credibility, increase costs and weaken morale. In many cases, the indirect consequences of poor quality, such as loss of trust or damaged relationships, are more severe than the immediate technical failure itself.

Quality is also important because it is closely linked to organisational performance. Reliable processes reduce variability, rework, and waste. Clear requirements and consistent delivery improve coordination and decision-making. Thus, quality is a prerequisite for efficiency and productivity. From a long-term perspective, quality is inseparable from organisational viability. An organisation that systematically fails to meet expectations will struggle to maintain customers, retain employees, or justify its role in society.

This is where the connection to sustainability emerges. Without quality, there cannot be sustained performance. Therefore, quality management provides an essential foundation for organisations that aim to be viable, resilient, and sustainable over time.

1.3 Expectations, requirements, and value

Quality becomes meaningful only when linked to expectations. Customers—whether individuals, organisations, or society at large—approach products and services with assumptions about what they should receive. These expectations may be explicit or implicit, precise or vague; however, they always exist. It is useful to distinguish between a few closely related concepts:

- Expectations reflect what customers believe or assume will be delivered. They are often shaped by prior experience, reputation, professionalism, and trust.
- Requirements are expectations that are expressed in a more explicit form. They may be stated verbally, written into contracts, or agreed upon during negotiations.
- Specifications translate the requirements into more detailed and verifiable descriptions. They define what is to be delivered and often how compliance can be assessed.
- Standards provide recognised reference points for defining requirements and specifications. They represent a shared agreement on good practice and can help align expectations between parties who may not know each other well.

A simple example illustrates the distinction between expectations, requirements, specifications, and standards. Consider a couple renovating the kitchen in their home. We can refer to the couple as two principal stakeholders. Their expectations may include that the kitchen is functional, visually pleasing, safe to use, and completed within a reasonable timeframe. Their expectations may be similar, but they are likely to differ in some ways. Some of these expectations are explicitly discussed, while others remain implicit. From these expectations, requirements are defined and agreed upon by the stakeholders. These may include the number and type of cabinets, appliances to be installed, delivery dates, and a target budget. The specifications then translate these requirements into detailed and measurable terms, such as exact dimensions, materials, technical solutions, and installation methods, typically documented in a contract.

Simultaneously, the project must comply with relevant standards and regulations, such as those in relation to electrical safety, ventilation, fire protection, and building codes. These standards apply regardless of individual preferences and ensure that the minimum levels of safety and performance are met. Quality is therefore assessed by how well the finished kitchen meets the couple's expectations, fulfils the agreed-upon requirements and specifications, and complies with applicable standards.

These distinctions are always important, especially in knowledge-based organisations such as consultancies, schools, hospitals, engineering firms, and other professional service providers. In such settings, customers are rarely able to specify all requirements in detail, and they can also have conflicting expectations. Instead, quality is strongly linked to professional judgement, competence, and trust. The customer expects the provider not only to follow stated requirements, but also to act responsibly when those requirements are incomplete or ambiguous. In this context, structured ways of working help ensure that expectations are interpreted consistently and fulfilled in a transparent manner.

Expectations and requirements are closely tied to quality features or characteristics, and quality is almost never defined by a single characteristic alone. Customers typically assess a combination of characteristics. Even a very simple product illustrates this point. At first glance, a pencil appears to be uncomplicated. However, it has several characteristics that may matter to the user: whether it has an eraser, whether the graphite breaks

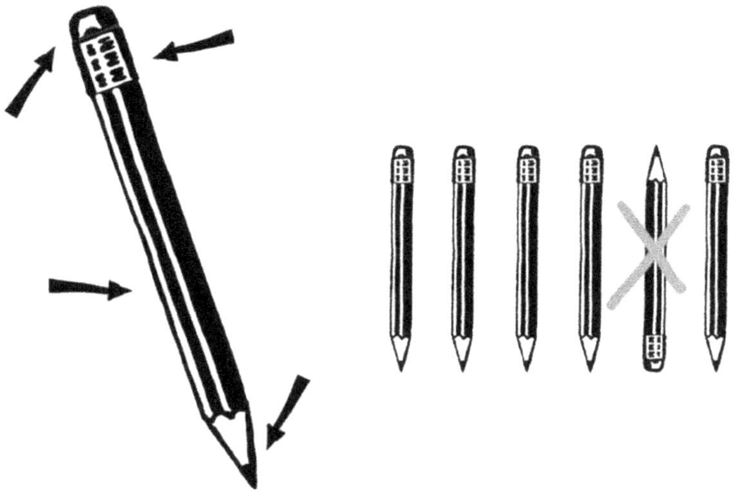

Figure 1.3 A simple product—and yet with specific characteristics.

easily, how smoothly it writes, and how comfortable it is to hold. None of these characteristics alone defines quality. Instead, quality emerges from the overall combination, relative to the user's expectations and intended use.

The same logic applies to services. A professional advisory service, for example, may be judged on accuracy, responsiveness, clarity of communication, availability, and ethical conduct. Some of these characteristics may even be conflicting, such as accuracy and responsiveness. From the customer's perspective, quality results from how well this bundle of characteristics aligns with expectations. This bundle can also be understood as the benefits the customer perceives in the product or service.

Some characteristics are inherent, meaning they are intrinsic to the product or service, such as reliability and durability. Others may be assigned, such as price. Price is one of several sacrifices the customer makes, which leads to the concept of value. In this book, value is introduced as a simple conceptual relationship:

$$\text{Value} \approx \text{benefits/sacrifices}$$

This expression is intended as a way of thinking rather than a precise mathematical formula. From a customer perspective, value reflects the relationship between perceived benefits and the sacrifices required to obtain them. We typically seek offerings that provide sufficient benefits relative to cost, time, effort, and other sacrifices. A product or service with strong benefits but excessive sacrifices may be perceived as poor value, just as an offering with low sacrifices that fails to deliver meaningful benefits will also be rejected.

At this point, it is useful to briefly return to the photograph discussed earlier in the chapter. When a client commissions a photograph, some expectations may be expressed as clearly defined requirements such as subject matter, resolution, or delivery format. Other expectations may be less explicit and relate to composition, mood, originality, or emotional impact. The photograph will therefore be judged not only against specified requirements, but also against broader characteristics that may be difficult to define fully in advance. Meeting technical requirements may be necessary, but perceived quality may ultimately depend on whether the final result fulfils both explicit and implicit expectations.

1.4 Quality, choice, and product categories

Discussions about quality often become confusing because products and services are compared across fundamentally different contexts. It is important to recognise that quality is closely linked to choice, expectations, and the product or service category within which that choice is made.

When my son was young, discussions about where to eat were usually brief. I might have suggested a restaurant with a particular atmosphere or culinary ambition, but he would almost always choose McDonald's. From his perspective, this choice was perfectly rational. He knew exactly what to expect: a familiar menu, predictable taste, short waiting times, and often a small toy as part of the experience. The food would look and taste the same regardless of the location, and the service would follow a recognisable pattern. For him, consistency and reliability define quality. My expectations were different, but his were consistently fulfilled—and in that sense, the restaurant delivered the "right" quality for its chosen category.

This example illustrates two important aspects. First, quality must be understood in relation to the product categories. Comparing fast-food restaurants with full-service restaurants using the same quality criteria is misleading. Within each category, quality can be meaningfully assessed; across categories, absolute comparisons such as "high" or "low" quality are of limited value.

Second, the example highlights two foundational dimensions of quality, as emphasised by Joseph Juran: quality of design and quality of conformance. Quality of design concerns what is intended, the features, specifications, and characteristics deliberately built into a product or service. Quality of conformance concerns how consistently actual delivery matches that intended design. In organisations such as McDonald's, standardisation is a deliberate design decision. The quality of the design is reflected in a carefully defined offering: menu items, preparation methods, service processes, layout, and customer interaction. Quality of conformance is reflected in how consistently this design is executed across locations and over time. For example, a McDonalds customer ordering a standard cheeseburger expects it to have the same taste, appearance, and presentation regardless of where and when it is purchased. Together, these two aspects strongly shape the customer's perception of quality.

The same applies to other contexts. Consider the example of a newly built apartment sold by a construction company. Customers typically expect certain design features, such as layout, materials, insulation level, and technical systems, to be consistent with specifications and marketing information. Simultaneously, they expect these features to be delivered without deficiencies. Installations should function properly, finishes should meet the agreed standards, and common areas should be completed as promised.

If the apartment has attractive features but is delivered with defects, customers will perceive the wrong quality. Flawless construction cannot compensate for a design that does not meet expectations or is misaligned with the intended market segment. Quality is therefore assessed through a combination of what is offered (quality of design) and how well it is delivered (quality of conformance).

Within a given product or service category, consistency and predictability are often decisive quality attributes. Customers value knowing what to expect and being able to rely on that expectation being met repeatedly. Variability that is not aligned with expectations is frequently perceived as poor quality, even if individual outcomes may appear acceptable in isolation.

This perspective reinforces a key message: quality is about designing offerings that fit clearly defined expectations and delivering them consistently through standardised and well-controlled processes.

Across these examples, one core lesson becomes clear: quality cannot be understood in a meaningful way without understanding the context. Different products, services, and situations emphasise different aspects of quality—technical reliability in one case, consistency and speed in another, creativity or experience in a third. For example, a customer visiting a standardised fast-food restaurant may value speed, predictability, and consistency, while a customer visiting a fine-dining restaurant may prioritise atmosphere, creativity, and personalised service. The management of quality must therefore always begin with understanding what matters in a given context, whose expectations define success, and which dimensions of quality are most important. This context-dependent perspective provides an important foundation and leads naturally to a broader systems view, where organisations must be understood not in isolation, but in relation to the environments, stakeholders, and interconnected processes that shape both expectations and outcomes.

1.5 A first-systems view of quality

Thus far, we have discussed quality in relation to products, services, expectations, and choice. To understand how quality can be managed in practice, we need to introduce a systems view.

At a basic level, a system can be understood as a set of interconnected elements that work together to achieve a specific purpose. The behaviour and performance of a system cannot be fully explained by examining its parts in isolation; they emerge from the interactions between those parts.

A useful distinction is often made between closed and open systems. A closed system is largely isolated from its environment and can be analysed primarily using internal variables. An open system, on the other hand, interacts continuously with its environment, exchanging resources, information, and value. Organisations are almost always open systems, even when they are managed as closed systems.

Consider the example of a hotel. At first glance, it may seem possible to manage it by focusing only on internal operations: reception, housekeeping, maintenance, food service, and administration. If hotels are treated as closed systems, quality management may focus narrowly on internal procedures and performance indicators.

However, the hotel's quality performance depends heavily on its interactions with the environment. Suppliers influence the availability and quality of food, linens, cleaning materials, and technical equipment. Guests arrive with expectations shaped by prior experiences, online reviews, price levels, and category positioning. Regulations influence safety, hygiene, and accessibility requirements. External events, such as seasonality, tourism patterns, and transportation disruptions, also affect performance.

Assuming the hotel to be a closed system can lead to quality problems. For example, a hotel may carefully design its service offerings by defining room standards, service levels, and amenities. However, it may still fail to meet expectations if supplier reliability is poor, staff turnover is high, or external communication is unclear. Quality issues may

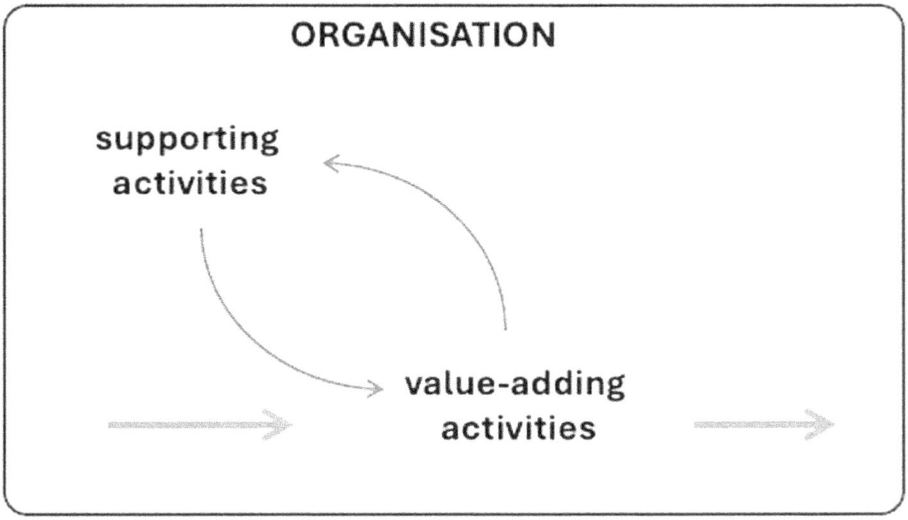

Figure 1.4 The organisation as a closed system.

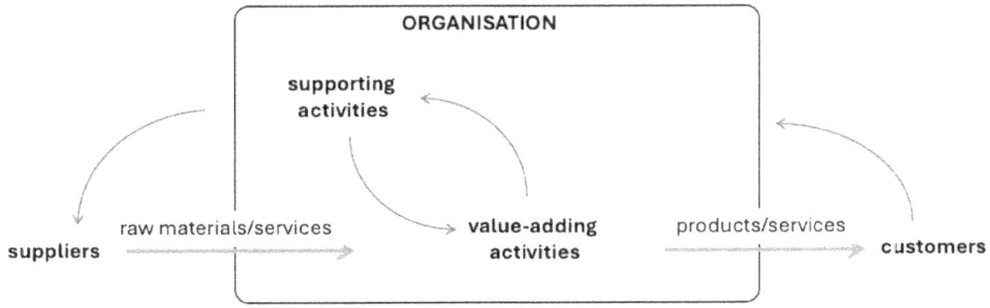

Figure 1.5 The organisation as an open system.

then appear "inside" the hotel, even though their root causes lie outside its immediate control.

This example also highlights the distinction between the quality of design and the quality of conformance, already introduced. The quality of design relates to decisions about the hotel's category and offering, room size, facilities, service concept, and target market. Quality of conformance relates to how reliably the design is delivered in daily operations, rooms cleaned as promised, services provided on time, and facilities functioning properly.

Different hotel categories are designed to meet different expectations of guests. A budget hotel and a luxury hotel do not compete on the same quality attributes, even though they provide the same basic service of accommodation. Therefore, quality must be assessed relative to category-specific expectations.

From a systems perspective, quality management in a hotel involves aligning design decisions, operational execution, and external interactions to ensure that expectations are consistently met. Improving housekeeping performance alone, for instance, may have a limited effect if upstream design decisions or supplier relationships are weak.

This system view highlights the need for a holistic perspective on quality management. This perspective is made more concrete through the concept of the value chain, which provides a structured way of understanding how quality is created or lost through interactions between suppliers, internal processes, and customers.

1.6 The value chain as an open system

We now turn to the value chain, which can be understood as an organisation viewed explicitly as an open system. An open system continuously interacts with its environment. It receives inputs, transforms them through internal processes, and delivers outputs that are evaluated by external actors. These inputs and outputs include physical materials, information, knowledge, expectations, requirements, and feedback.

The value chain concept is commonly associated with Michael E. Porter (1985),[2] who introduced it as a way of analysing how organisations create value and gain competitive advantage. From a quality perspective, the value chain helps us understand that quality is not created at one point. It emerges from interactions between suppliers, the organisation's internal processes, and customers.

As an open system, the organisation depends on suppliers for materials, components, services, and expertise. These inputs are transformed through a set of internal activities

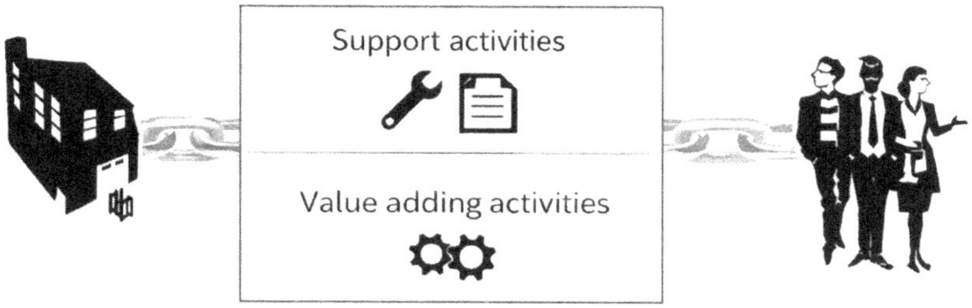

Figure 1.6 The value chain, linking an organisation to its suppliers and customers.

and processes that apply knowledge, technology and human effort. The outputs are products or services delivered to customers who then assess whether their expectations have been met. Feedback from customers, regulators, and other stakeholders flows back into the system and influences future decisions and performance.

This perspective has several important implications. Quality problems rarely originate in isolation. They may stem from unclear expectations, weak interfaces between processes, poor supplier performance, or inadequate feedback mechanisms. Similarly, quality improvements often require changes that extend beyond a single department or function. Seeing the value chain as an open system helps explain why local optimisation can be ineffective, and why coordination and alignment across boundaries are so important.

The open-systems view also draws attention to the relationships and cooperation within the value chain. In many project-based and complex organisational settings, closer collaboration between clients, designers, contractors, operators, and suppliers can improve the understanding of requirements and reduce downstream problems. Investing effort early in clarifying expectations and interfaces often leads to better outcomes and lower overall costs over the life cycle.

In the case of a hotel, value is created through a chain that includes suppliers of food, linen, cleaning services, furniture, energy, and digital booking systems. Internal processes, such as reception, housekeeping, and facility management, transform these inputs into a service experience that guests evaluate against their expectations. Quality problems observed by guests often originate upstream, for example, in design decisions, supplier performance, or staffing arrangements, rather than at the point of direct interaction.

A similar logic applies to fast-food restaurants. Here, the value chain is deliberately designed to be highly standardised. Suppliers, processes, and interfaces are specified to ensure consistency across locations. Both the quality of design (the standardised offering and processes) and the quality of conformance (the reliability of execution) are embedded throughout the value chain rather than concentrated at the point of sale.

The same open-system logic is evident in the case of newly built apartments. Value creation begins long before a buyer takes possession. Architects, engineers, contractors, material suppliers, and regulatory bodies contribute to shaping the final outcome. The buyer's perception of quality reflects both design features and conformance. Design features can include layout, materials, energy performance, and technical solutions. Conformance refers to whether these features are delivered as specified. Defects experienced at handover often reflect misalignment earlier in the value chain, rather than isolated workmanship issues.

In construction and real estate development, this insight has led to the increased use of partnering arrangements. Partnering refers to a form of collaboration in which key actors in the value chain, such as designers, contractors, and operators, are involved early and work together under shared objectives. Rather than treating each actor as an isolated contractual counterparty, partnering recognises their interdependence and seeks to align incentives, responsibilities, and information flows. Experience shows that such arrangements can reduce rework, improve quality outcomes, and lower lifecycle costs by addressing quality issues at their source, rather than after delivery.

The value chain perspective also reinforces the importance of aligning design decisions and execution across organisational boundaries. Quality problems are more likely to occur when design choices, supplier capabilities, and operational practices are not aligned. When the value chain is understood and managed as an integrated system, organisations are better able to consistently meet expectations.

Finally, it is important to note that real-world value creation often extends beyond a simple, linear chain. In many contexts, organisations operate within value networks, where multiple actors interact in non-linear ways, exchanging resources, information and services across formal and informal boundaries. Digital platforms, service ecosystems, and long-term partnerships are examples of such networks. Although the value chain provides a useful starting point, the concept of value networks highlights the growing complexity of quality management in interconnected systems.

1.7 Quality assurance: Creating confidence

Quality assurance involves creating confidence that quality requirements will be fulfilled. From the perspective of customers and other stakeholders, it answers a simple question: can I trust that this organisation will deliver what it promises?

In everyday life, people rely on many different signals to form this judgement. Reputation is one of the strongest forms of quality assurance. A simple and well-known example is Bæjarins bestu, a small hot dog stand in central Reykjavík. The operation itself is simple, but it has consistently provided its services for decades. As a result, people expect a certain taste, price level, and service experience. The long queue that often forms is not accidental; it reflects the accumulated trust. Customers rely on reputation built through the repeated fulfilment of expectations.

Another familiar form of quality assurance is the use of guarantees and return policies. A well-known example is IKEA. Customers know that if a product does not meet their expectations, it can be returned within a generous timeframe, and a refund will be provided. This reduces the perceived purchase risk and signals that the company is confident in its products. Simultaneously, such policies create strong internal incentives to prevent defects and misunderstandings.

Quality assurance can also be created through clear and transparent procedures that make it easy for customers to understand how a service operates. A good example is Uber. Customers know in advance how the service is ordered, how the pricing is calculated, and how the payment is handled. During the service, the driver's approach can be followed in real time on a digital map. This transparency reduces uncertainty and builds confidence even before the service is delivered.

A further step in quality assurance is when organisations explicitly refer to recognised standards in their operations. For example, a consulting firm may state that its management system is based on an international quality management standard, or a

Figure 1.7 How to assure the customer that their expectations will be met?

medical laboratory may refer to a recognised accreditation standard for testing and calibration. Such references signal that the organisation's way of working is structured and aligned with accepted professional practice rather than being improvised or purely individual.

Quality assurance can be further strengthened through independent certification, where an external party assesses whether the organisation's practices conform to a specified standard. Many organisations claim that they are certified according to standards such as the ISO 9001. What this actually means—and what it does not mean—will be discussed in later chapters. At this stage, it is sufficient to note that certification functions as an institutional trust mechanism, particularly in situations where customers have limited direct experience or where the consequences of failure are significant.

Taken together, these examples illustrate that quality assurance exists on a continuum, from informal trust based on reputation, through tangible risk-reducing mechanisms and transparency, to formal standards and independent certification. In all cases, the purpose is the same: to reduce uncertainty, build confidence, and support stable relationships between organisations and their stakeholders.

1.8 Quality cost and the economics of quality

So far, quality has been discussed mainly as a matter of meeting expectations and delivering value. The next question is how cost relates to quality. Quality itself is not a cost in the same way as materials, labour, or equipment. What is costly, however, is the failure to deliver the right quality.

When a product or service fails to meet expectations, the consequences are often expensive. These costs may take many forms, e.g. rework, delays, complaints, warranty claims, loss of reputation, loss of trust, or even the loss of future business. From the customer's perspective, poor quality represents wasted time, frustration, and uncertainty. From the organisation's perspective, it represents inefficiency and lost value.

Quality cost can be defined as the total cost incurred by an organisation to ensure quality, as well as the costs resulting from failures in systems, processes, products, or services. These costs occur throughout the organisation, in sales, design, service delivery, administration and customer interaction. They are influenced not only by internal activities but also by suppliers, partners, and customers.

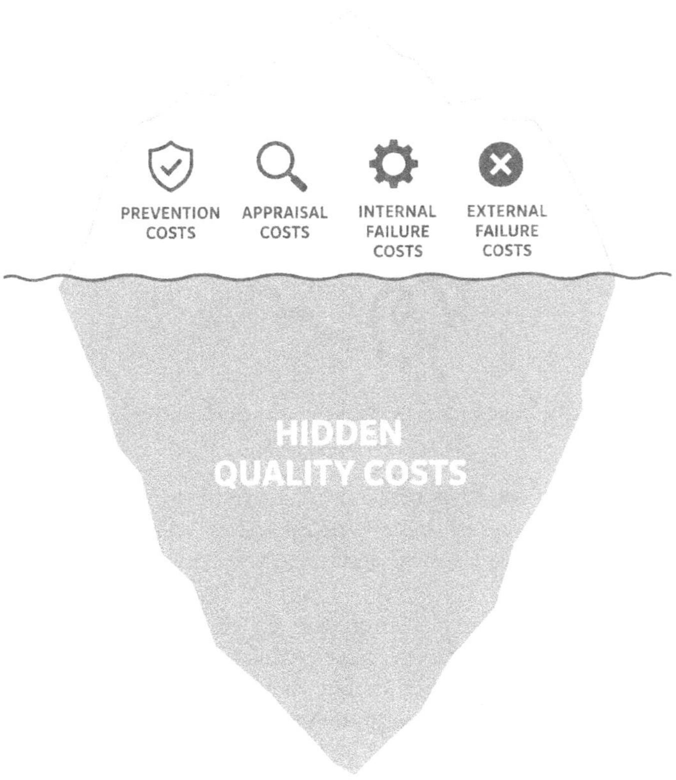

Figure 1.8 Categories of quality cost.

To illustrate the different categories of quality cost, consider a professional accounting firm that provides bookkeeping and annual financial reports for small and medium-sized enterprises.

Failure costs

Quality failures can be classified into internal and external failures.

Internal failure costs arise when errors are discovered before a service is delivered to a client. For example, an accountant may discover mistakes in a financial report during an internal review process. Correcting these errors typically requires extra work, leads to delays, and consumes staff time. Although the client may never become aware of the problem, the organisation still bears the cost.

External failure costs arise when errors are discovered after delivery. If a client identifies mistakes in an annual report, the accounting firm must correct the work, hold additional meetings, reissue documents, and possibly explain the error to the authorities. Beyond the immediate cost of rework, such failures can damage trust and reduce the likelihood of future business.

In general, external failures are more costly than internal failures because they involve not only correction but also relationship management and loss of confidence.

Appraisal costs

Appraisal costs are incurred to detect errors before the delivery. In an accounting firm, this may involve having a senior accountant review all financial reports before they are sent to clients. Although such reviews help reduce external failures, they require significant time and expertise. Therefore, excessive reliance on appraisals can be costly and inefficient.

Prevention costs

Prevention costs are incurred to avoid errors occurring in the first place. In accounting firms, prevention may include staff training, development of clear work procedures, use of standard templates, and continuous improvement of internal processes. These activities require an upfront investment but reduce the likelihood of internal and external failures.

From a quality management perspective, prevention is usually the most effective way to reduce the total quality cost.

Hidden costs

In addition to visible costs, hidden costs must be recognised as a distinct category. These include loss of reputation, reduced customer loyalty, increased employee stress, time spent firefighting, and management attention diverted from development to problem-solving. Hidden costs are often difficult to measure, but they can be substantial.

The application of quality management should lead to a reduction in the total quality cost. Increasing prevention costs may appear expensive in the short term, but they typically lead to significant reductions in failure and appraisal costs, resulting in lower overall costs and higher customer satisfaction.

By expressing quality-related problems in financial terms, quality cost makes the consequences of poor quality visible and tangible. This helps shift quality management from being perceived as a "nice to have" to a "need to have"—a necessary condition for efficient operations, satisfied customers, and long-term value creation.

Ultimately, organisations invest in quality management because poor quality is expensive, and in many ways, quality management is fundamentally about improving business performance. Defects, rework, delays, complaints, lost customers, and damaged reputation all have financial consequences. Therefore, effective quality management should lead to lower operating costs, more stable revenues, and stronger long-term profitability.

1.9 Quality, sustainability, and long-term value creation

In this chapter, quality has been presented as a concept rooted in expectations, value creation, and systems thinking. These same ideas form the foundation of sustainability, understood here as an organisation's ability to create value over time within economic, social, and environmental constraints.

From this perspective, quality management is a prerequisite of sustainability. Organisations that systematically fail to meet expectations waste resources, erode trust, and weaken their viability. Rework, defects, complaints, and firefighting consume time, materials, energy, and human effort that could otherwise be used more productively.

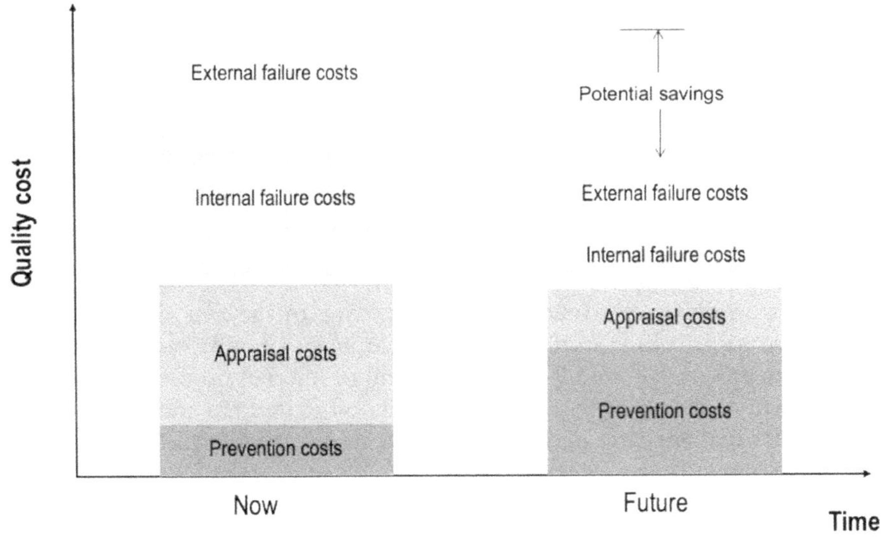

Figure 1.9 Quality costs and potential savings.

The concept of quality cost, discussed in the previous section, makes this connection explicit. Internal and external failures, as well as hidden costs such as loss of reputation or employee frustration, reflect the value that is destroyed rather than created. Reducing such losses through prevention and learning improves economic performance and resource efficiency. Therefore, quality improvement supports sustainability by reducing waste and increasing the effectiveness of value creation.

Quality is also closely linked to organisational learning and resilience. Organisations that systematically manage quality can detect problems earlier, analyse their causes, and adapt their processes accordingly. This learning-oriented approach strengthens the organisation's ability to cope with change, whether driven by shifting customer expectations, technological development, regulatory pressure, or broader societal demands. Resilient organisations are better positioned to remain viable in the long term. In turbulent environments, structured quality management provides stability without preventing adaptation.

From a systems perspective, sustainability depends on maintaining a balance within and across value chains and value networks. Quality management contributes to this balance by aligning expectations, design choices, execution, and feedback among interconnected actors. When quality is managed narrowly or locally, improvements in one area may cause problems elsewhere. When quality is managed systemically, value creation can progress more smoothly without transferring costs or risks to other stakeholders or future periods.

Sustainability does not imply maximising quality in all dimensions. Just as quality is defined relative to expectations and context, sustainable performance requires conscious trade-offs. Quality management provides a structured way to make these trade-offs visible and manageable rather than implicit and reactive. The field expands from serving immediate customer needs to supporting long-term stakeholder value within societal and environmental boundaries.

This chapter has established quality as a foundational concept for understanding how organisations create, deliver, and sustain value. The chapters that follow build on this foundation by exploring how quality management has evolved as a discipline, how it is formalised through systems and standards, and how it can be implemented and developed in practice.

Case: Rising costs in a house painting company

A medium-sized house painting company specialises in the indoor painting and decorating of newly built apartments and small commercial buildings. The company employs around 20 people, most of whom are trained painters with several years of practical experience. Management sees this strong professional competence as one of the company's key strengths.

The company has traditionally enjoyed a good reputation and has been able to secure a stable flow of work. Revenues have remained fairly stable in recent years, although competition has increased, and prices have come under pressure. Simultaneously, profitability has steadily declined.

A few years ago, the company completed a large apartment project that resulted in significant customer complaints. Some apartments needed to be repainted because of uneven finishes, poor surface preparation, and inconsistent colour matching. Teams had to return to the site several times after the handover. Management believes that this project may also have damaged the company's reputation and reduced the likelihood of repeat business.

In response, the company increased its inspection and checking activities. More supervisors became involved in reviewing completed work before delivery, and painters were increasingly asked to redo minor details before handover. This reduced the number of external complaints but also increased the amount of time spent on supervision, coordination, and corrections.

The company does have a simple quality manual, but it is rarely used in practice and is mainly seen as something that exists "for show." New employees learn mostly through experience and informal guidance from their colleagues. There is little formal training, limited use of checklists, and few common work procedures.

Management is beginning to question whether the current approach is sustainable. Complaints have decreased, but profitability continues to decline.

The table shows a simplified income statement (all amounts are in thousands of EUR).

Table 1.1 Income statement of a house painting company

	Present year	*Last year*
Revenue	2720	2900
Direct labour cost	-1420	-1280
Materials and equipment	-560	-500
Staff training and work procedure development	-40	-30
Inspection, supervision, and final checks	-150	-110
Warranty work and customer complaint handling	-130	-210
Administration and overhead	-280	-270
Operating cost	-2580	-2400
Operating profit	140	500

If you were advising the managing director of this company, what would you recommend? Which underlying problems would you focus on first, and what changes could help improve both quality and profitability in the longer term?

Reflection Points—Chapter 1

1. When "premium quality" becomes meaningless
   A company may proudly market its product as "premium quality," yet customer dissatisfaction may increase and returns may rise. Analyse this situation using the distinction between the quality of design and quality of conformance. Is the problem related to expectations, specifications, execution or market positioning? How might vague language around "high quality" obscure the real managerial issue?
2. Managing conflicting expectations
   Public hospitals must simultaneously satisfy patient expectations, regulatory requirements, budget constraints, professional standards, and political scrutiny. Using the concepts of expectations, requirements, and stakeholders, consider how quality should be defined in a healthcare context. Whose expectations should take precedence when they conflict, and what risks arise if stakeholder expectations are ignored?
3. Fit for purpose
   A technology company has continuously added new features to its products to differentiate itself in the market; the result, however, has been increased complexity, a decline in reliability, and rising customer support costs. Has the quality improved or deteriorated? Apply the concept of product categories and fit-for-purpose to evaluate whether adding features necessarily enhances value.
4. The cost of control
   A consulting firm frequently detects and corrects internal errors before reports reach clients, and management views this as evidence of strong quality control. Using the logic of quality cost, analyse whether this situation may still represent a strategic risk. At what point does internal failure become a systemic weakness rather than a manageable inconvenience?
5. The illusion of the closed organisation
   A hotel chain focuses exclusively on improving internal efficiency and standardising procedures but pays little attention to supplier reliability, labour conditions or external perceptions. Using the systems perspective and the concept of the value chain, explain how quality may be created or lost outside the organisation's direct boundaries. What risks arise from managing quality as if the organisation were a closed system?
6. Prevention as strategic investment
   A manufacturing company reduces training budgets and preventive activities to improve short-term financial performance. Initially, the reported costs decline, and profitability increases. Using the concept of quality cost, analyse the likely long-term consequences of this decision and discuss whether apparent savings may conceal future risks.

Notes

1 Juran, J. (1989). *Juran on Leadership for Quality: An Executive Handbook*. New York: Free Press.
2 Porter, M.E. (1985). *Competitive Advantage: Creating and Sustaining Superior Performance*. New York: Free Press.

2 The roots and evolution of the field, and how it relates to sustainability

2.1 Setting the scene: Origins and evolution

Quality management is often introduced through tools, techniques, and standards. While these are important, they make much more sense in the context of how the field emerged and why particular ideas took shape. Therefore, this chapter frames quality management as an evolving field of thought and practice. It is, in fact, a field shaped by growing organisational complexity, global connectivity, technological development, and expanding expectations regarding responsibility and long-term value creation.

Many foundational ideas associated with quality management were developed by scholars and practitioners in the United States. Concepts such as statistical thinking, systematic management, and structured improvement were articulated by a small group of influential thinkers. However, the environment in which these ideas were most fully embraced and developed into everyday practices was post-war Japan. In the decades following World War II, quality-oriented thinking became deeply embedded in Japanese industry, supported by strong managerial commitment, collective learning, and a culture that valued continuous improvement.

This historical pattern helps explain why quality management is not merely a collection of methods but also a question of organisational culture, leadership, and shared responsibility.

Figure 2.1 Evolution of quality management.

DOI: 10.4324/9781003671749-2

This chapter highlights some key contributors and ideas that have proven durable and influential. The emphasis is on how quality management expanded in scope from the inspection of products to the control of processes to organisation-wide management responsibility and ultimately to integrated approaches concerned with long-term performance, risk, and stakeholder value.

2.2 Early managerial roots, from craft to scientific management

Before the emergence of quality management as a distinct field, quality was largely ensured through craftsmanship. In craft-based production, such as that of carpenters, the same individual is responsible for designing, producing, and delivering the product. The craftsperson worked directly for a specific customer, understood the requirements, and bore full responsibility for whether the result met the expectations. Quality was personal, visible, and closely tied to their professional pride.

This situation fundamentally changed with industrialisation. As production volumes increased and work was divided among many individuals, the direct link between the worker, product, and customer weakened. Tasks were specialised, coordination became more complex, and responsibility for quality could no longer rest with a single individual. Instead, quality gradually became an organisational concern, managed through supervision, inspection, and formal roles.

An important influence on this transformation was Frederick Winslow Taylor, whose book *The Principles of Scientific Management* (1911)[1] laid the foundations for management as a scientific discipline. Although Taylor is not usually regarded as a quality guru, his contribution is central to the later development of quality management. One of his key arguments was that managers are not born but can and should be trained—an idea that established management as a professional field based on systematic analysis and education.

A core principle of Taylor's thinking was the separation of planning and execution. Managers were responsible for analysing work, defining standard methods, and setting expectations, whereas workers were responsible for carrying out the defined tasks. This shift made work more predictable and measurable, but it also transformed the way quality was managed. Once production was organised through central planning and standardisation, quality could no longer rely solely on individual craftsmanship. Instead,

Figure 2.2 The craftsman—early managerial roots.

Figure 2.3 A foreman with his team of carpenters.

inspection and control became formal responsibilities within the organisation, and dedicated quality functions began to emerge within it.

While scientific management has brought discipline, efficiency, and repeatability, it has also revealed important limitations. Taylor's ideas later became known as Taylorism, a term often used to describe highly standardised, efficiency-driven forms of work organisation. Over time, Taylorism was criticised for being dehumanising to workers and neglecting the social and psychological dimensions of work. Critics have argued that reducing work to predefined tasks and numerical targets risks treating people as extensions of machines rather than as thinking contributors.

These criticisms are important, but Taylor's fundamental contribution is also important: establishing management as a discipline that can be studied, taught, and improved. Many of the limitations associated with Taylorism were later addressed—explicitly and implicitly by quality management thinkers, who reintroduced learning, cooperation, and human judgement into systematic management.

Figure 2.4 An inspector, monitoring the output of the workers.

2.3 Statistical thinking and the logic of variation: Shewhart

The limitations of inspection-based control and scientific management created the conditions for a decisive conceptual breakthrough in quality management, that is, the recognition that quality problems are fundamentally linked to variation in processes. This shift is most clearly associated with the work of Walter A. Shewhart, whose ideas form the intellectual foundation of modern quality management (Shewhart, 1931).[2]

Shewhart approached quality from both technical and conceptual perspectives. He argued that quality has two inseparable dimensions. On the one hand, there is the subjective side, which includes what customers need, expect, or value. On the other hand, there is the objective side, the measurable characteristics of a product or service that can be observed, analysed, and controlled. In his view, quality management must address both dimensions simultaneously.

A central contribution of Shewhart was his insight that all processes exhibit variation, and that not all variation should be treated in the same way. He distinguished between two different types of variation. Common causes are inherent to the process itself and reflect its normal behaviour. These cannot be eliminated through local fixes or inspections; they require changes in the process design or operating conditions. In contrast, special causes arise from specific, identifiable events that are external to the normal process, such as equipment failures, unusual inputs, or human error. These issues can and should be detected and addressed directly.

The idea of variation can be illustrated with something most people recognise, the download speed of a home internet connection. Even when a provider promises a certain speed, the actual speed fluctuates during normal use. When these fluctuations are small and the connection feels stable, users rarely notice them. This represents a common variation in the service. Occasionally, however, the connection becomes unusually slow or stops working altogether. In such cases, users immediately sense that something is wrong and expect action to be taken. This represents a specific variation caused by an identifiable problem that lies outside normal operations.

From the user perspective, quality is associated with stable and acceptable performance. Large or frequent fluctuations indicate a quality problem. Therefore, understanding whether instability reflects normal process behaviour or a special cause is central to managing quality. This distinction has profound implications. This shows that reacting

Figure 2.5 A controller with his control chart.

to every deviation as if it were a problem is ineffective. Instead, managers must learn to understand process behaviour statistically so that they can respond appropriately, stabilise processes by removing special causes, and improve them by redesigning the system itself.

To support this way of thinking, Shewhart developed methods that later became known as statistical process control (SPC). Control charts made it possible to monitor processes over time, distinguish between normal variation and signals of abnormal behaviour, and base decisions on data rather than intuition or blame. In this sense, Shewhart moved quality management away from after-the-fact inspection towards process understanding and prevention.

Closely related to this statistical logic is the iterative management cycle, often referred to as the plan–do–check–act (PDCA) cycle. Although it was later popularised by others, the underlying logic originated with Shewhart. The cycle reflects a disciplined approach to learning: planning an intervention, executing it, studying the results, and acting on what is learned. This simple but powerful idea later became central to continuous improvement and organisation-wide quality management.

Shewhart's work marked a turning point in the evolution of quality management. It introduced a way of thinking that connected data, learning, and management responsibility and provided the conceptual tools that later contributors would build upon. Quality management was no longer primarily about inspection or compliance, but about understanding and improving processes over time.

2.4 Quality as a management philosophy: Deming

W. Edwards Deming transformed Shewhart's ideas into a comprehensive and deeply critical philosophy of management. Deming's message was radical for its time; improving quality required nothing less than a fundamental transformation of management practice.

Deming insisted that quality must be defined by the customer. Quality, in his view, could not be reduced to internal specifications or technical conformance alone but had to be understood in terms of customer satisfaction. He emphasised that customers are diverse, and what satisfies one customer may not satisfy another. The continued existence of every organisation depends on delivering products and services that satisfy its customers.

Deming articulated his ideas in his book Out of the Crisis[3] (Deming, 1982). The "crisis" he referred to was the very real loss of competitiveness experienced by the American industry, particularly in comparison with Japanese firms, for example, in the automotive sector. Deming argued that this decline could not be explained by labour costs or technology alone but was rooted in management practices such as short-term thinking, excessive reliance on inspection, management by targets, and a failure to understand variation and learning. Deming warned against reducing organisations to spreadsheets and numerical targets. He argued that such approaches ignore the systemic nature of work and undermine learning, cooperation and intrinsic motivation. Barriers between departments should be removed, and other departments should be seen as internal customers. Deming's ideas are clearly articulated in his well-known 14 points for management, forming an integrated argument for a different management philosophy.

The plan–do–check–act (PDCA) cycle provides a practical expression of this philosophy of continuous improvement. Deming elevated PDCA to a general model for

Figure 2.6 A specialist seeking to improve a process using the PDCA cycle.

management and organisational learning. This cycle reflects a disciplined approach to improvement. Planning changes based on an understanding of the system, deliberately implementing them, studying the results in relation to expectations, and acting on what is learned. When used properly, the PDCA discourages reactive decision-making and reinforces learning over time.

Deming's ideas repositioned quality management as a customer-oriented, system-focused, and learning-driven approach to management. Quality is the outcome of thoughtful leadership, informed by data, grounded in collaboration, and oriented towards long-term capability. This perspective strongly influenced later developments in quality management and provided a conceptual bridge to ideas such as quality culture, continuous improvement, and integrated management systems.

2.5 Defining quality management: Juran and the trilogy

Joseph M. Juran (1989)[4] made quality management more accessible to managers by framing it as a set of deliberate, manageable processes. He argued that quality is not confined to production or operations. He illustrated this by distinguishing between q, which refers to quality in production processes, and Q, which refers to quality across the entire organisation, including planning, administration, support functions, and management decision-making.

Juran argued that many quality problems cannot be solved by technical means alone but require leadership, education, and structured improvement. His most widely known contribution is his quality trilogy, which describes quality management as consisting of three interrelated managerial challenges: quality planning, quality control, and quality improvement. Quality planning focuses on identifying customers, understanding their needs, and designing processes that can meet those needs. Quality control involves evaluating actual performance, comparing it with planned objectives, and responding to deviations. Quality improvement addresses the need to reduce chronic waste and variation by deliberately changing processes.

If no changes are made to a process, it will continue to generate a certain level of waste and inefficiency. When improvement efforts are initiated, some costs are incurred in the short term due to analysis, training, redesign, and implementation. However, if the improvement is successful, the process stabilises at a new and better level of performance. Over time, the initial investment is recouped through reduced waste, lower variation, and improved margins.

This logic can be illustrated using a practical example. Consider a supermarket that competes by consistently offering customers low prices. Such a strategy often depends on high efficiency, careful planning, and tight control of operating costs, since profit margins on individual products may be limited. One key challenge is managing food waste, for instance, the quantity of fruit discarded daily because too much was purchased and too little sold. Excessive waste increases costs directly and may ultimately undermine the supermarket's ability to maintain competitive prices. Waste therefore becomes not only a cost issue and a sustainability concern but also a strategic quality parameter.

Through quality control, the supermarket monitors daily waste levels and compares them with planned expectations. If the waste remains within the expected range, no corrective action is required. However, if there is a sudden and significant increase in discarded fruit, for example, due to a refrigeration failure, delivery error, or unexpected disruption, this represents a specific variation. In such cases, the appropriate response is to identify and remove the specific cause and restore normal operation rather than redesign the entire process.

If management decides that the overall level of waste is too high, even when no abnormal events occur, it signals a different type of problem. The issue is no longer a specific variation, but rather chronic waste built into the process itself. Reducing waste requires quality improvement rather than tighter control. The supermarket may need to redesign its forecasting methods, change supplier agreements, adjust ordering frequencies, introduce dynamic pricing, or modify store layout and handling procedures.

Such changes are rarely achieved through day-to-day management alone. They require structured analyses, coordinated decision-making, testing of alternatives, and implementation over time. Therefore, quality improvement is often organised as an improvement project with defined objectives, allocated resources, responsibilities, and a limited timeframe. In this sense, Juran's trilogy provides an early and natural link between quality

Figure 2.7 The quality trilogy by Joseph Juran.

management and project management, a connection that will be explored further in this book.

2.6 System-wide quality and the cost perspective: Feigenbaum

While Juran provided managers with a clear framework for planning, controlling, and improving quality, Armand V. Feigenbaum expanded the scope of quality management to the entire organisation and its value chain. His central argument was that quality cannot be assured by isolated functions or departments; it is the result of how all parts of an organisation work together.

Feigenbaum (1991)[5] introduced the concept of total quality control, emphasising that quality is influenced not only by production activities, but also by design, procurement, logistics, coordination with suppliers, and interactions with customers. In this view, quality problems often originate far from the point where they become visible. A defect detected during production may have its root cause in design decisions, supplier performance, or inadequate communication between organisational units. For example, a construction company renovating kitchens in private homes may consistently deliver projects that are late and over budget. While problems become visible during execution, their root causes often lie in earlier phases, such as unrealistic planning, incomplete design decisions, or poor coordination with suppliers, rather than in craftsmanship on site.

A defining feature of Feigenbaum's contribution is his explicit focus on the economic dimension of quality and his definition of the cost of quality. He argued that quality must be understood and managed in terms of cost, not merely technical performance. When processes and systems do not consistently meet defined expectations, costs arise throughout the organisation, for example, through waste, rework, delays, complaints, loss of trust, and reduced future business. At the same time, efforts to prevent such outcomes, improve coordination, and redesign processes also require resources. Therefore, the task of management is to understand and optimise the total cost associated with meeting expectations.

This perspective reinforces the idea that quality is a strategic management issue. Decisions about suppliers, product design, process configuration, and organisational structure all have long-term quality and cost implications. Therefore, responsibility for quality must be embedded in managerial decision-making at all levels; it cannot be delegated to specialists or inspectors.

Feigenbaum also stressed that quality is ultimately defined by the customer's experience and perception, not by internal technical criteria alone. This reinforces earlier insights from Shewhart, Deming, and Juran, while placing them in a system-wide and economically grounded context. Quality is a competitive factor that influences customer choice, loyalty, and long-term organisational success.

2.7 Prevention, discipline, and "quality is free": Crosby

The central claim of Philip B. Crosby (1980)[6] was simple and deliberately provocative; "Quality is free." He did not mean that quality requires no effort, but that the cost of ensuring that requirements are met is typically lower than the cost of failing to meet them, including rework, waste, delays, and loss of trust.

Crosby defined quality as conformity to requirements. In his view, quality cannot be vague or subjective if it is to be managed effectively; requirements must be clearly defined and communicated to all stakeholders. Once this is done, quality becomes a matter of ensuring that the work is done correctly the first time. Deviations from the requirements are failures of the system—not inevitable nuisances.

A key element of Crosby's thinking was his strong emphasis on prevention rather than on inspection. He argued that organisations that rely heavily on inspection and rework have already accepted failure as normal. Inspection may detect defects, but it does not eliminate their causes. Preventive action through clear requirements, disciplined processes, training, and accountability is therefore both more effective and economical in the long run.

Crosby is also closely associated with the concept of zero defects. Crosby's intention was not to deny the existence of variation but to challenge the mindset that defects are acceptable or unavoidable. Zero defects represent a managerial standard and cultural aspiration, signalling that producing defects is not an acceptable outcome of normal operations.

Crosby reinforced and sharpened earlier ideas by explicitly linking poor quality to failure costs. These include internal failures such as waste, rework, and delays, as well as external failures such as complaints, returns, lost customers, and reputational damage. Crosby argued that when these costs are properly understood and made visible, investments in prevention and improvement are not expenses but sound economic decisions.

Crosby's claim that quality is free can be illustrated with a simple example. Consider a craftsperson hired to renovate an old chair. The client wants a straightforward restoration; functional repairs, new upholstery, and a specific colour. Instead, the craftsperson decides to add decorative features and redesign elements that were never requested. Although the final product may appear more elaborate, it does not conform to the client's actual requirements, as the work goes beyond what was agreed. If the customer rejects the result, the extra work creates waste rather than value, and the chair may need to be redone. The key point is that quality is not defined by doing more, but by meeting requirements correctly the first time. Costs arise not because quality is expensive, but because failing to deliver what was required creates unnecessary effort, rework, and lost value.

Crosby's contribution is particularly relevant for managers because of its clarity and directness. His work also reinforces the idea that quality is not primarily a technical issue, but a matter of organisational discipline, leadership commitment, and shared expectations.

Another important aspect of Crosby's contribution is his early articulation of quality management maturity. In Quality Is Free, Crosby described how organisations progress through identifiable stages in their approach to quality—from reactive inspection and firefighting, through increasing discipline and prevention, to more mature, systematic, and proactive quality management. This idea, that organisational capability in quality develops over time, proved highly influential. It later became the foundation for maturity models in fields such as software engineering and, eventually, project management.

Together with the contributions of Shewhart, Deming, Juran, and Feigenbaum, Crosby's ideas helped consolidate quality management as a coherent and effective managerial discipline. They also prepared the ground for approaches that emphasise continuous improvement, built-in quality, and systematic waste reduction—developments that would later be strongly associated with Lean thinking.

2.8 Continuous improvement and Lean as part of the evolution

The ideas introduced by Shewhart, Deming, Juran, Feigenbaum, and Crosby converged and evolved into approaches that emphasised continuous improvement, waste reduction, and built-in quality. This development is most clearly associated with what later became known as Lean thinking.

The practical embodiment of these ideas can be seen in the Toyota Production System (TPS). Developed in the context of post-war Japanese manufacturing, the TPS emerged through experimentation and learning within Toyota under the leadership of figures such as Taiichi Ohno and Eiji Toyoda. The system integrates statistical thinking, prevention, continuous improvement, and human engagement into a coherent production and management approach. Rather than relying on inspection and rework, TPS emphasised built-in quality, stable processes, and learning through systematic problem-solving.

Lean later emerged as a broader interpretation of these principles, applied consistently over time. Central to Lean is the recognition that value is defined by the customer and that activities that do not contribute to customer value represent some form of waste. Therefore, reducing waste improves quality, lowers costs, and increases responsiveness.

Several Japanese scholars and practitioners have contributed to refining and disseminating these ideas. Among them were Shigeo Shingo, Kaoru Ishikawa, and Yoshio Kondo, whose work emphasised practical problem-solving, employee involvement, and the importance of motivation and learning in improvement efforts. Their contributions reinforce the view that quality should be built into processes and supported by the active participation of those performing the work.

Lean thinking also reflects the earlier insights of Deming and Juran. Stable processes are required before meaningful improvement can occur, and improvement is understood as a continuous, iterative activity rather than a one-time intervention. In this sense, Lean can be seen as the operationalisation of the PDCA logic at scale, embedded into everyday work practices.

A defining feature of Lean is its strong focus on flow—ensuring that work progresses smoothly through processes without unnecessary delays, interruptions, or rework. Achieving flow requires attention to process design, coordination across functions,

Figure 2.8 Continuous improvement.

and close collaboration with suppliers. These concerns reinforce the idea that quality is shaped across organisational boundaries.

Equally important is the concept of respect for people. Lean emphasises that sustainable improvement depends on employees' engagement, competence, and initiative. Those who perform the work are seen as a critical source of knowledge about problems and opportunities for improvement. This emphasis links Lean directly to the idea of quality culture, where shared values, norms, and behaviours support continuous improvement.

Therefore, lean is best understood as a continuation of the evolution of quality management. It integrates statistical thinking, managerial responsibility, economic logic, and human engagement into a coherent approach that focuses on learning and improvement. This perspective makes it easier to understand Lean as an organisational capability developed over time.

2.9 From quality management to sustainability-integrated management

Over time, the scope of quality management has expanded. What began as a concern with inspection and control gradually evolved into a system-oriented and people-centred management philosophy. An increasingly important step in this evolution is the integration of quality thinking with broader concerns regarding long-term performance, risk, and sustainability. The broader perspective of sustainability is reflected in the United Nations Sustainable Development Goals (United Nations, 2015),[7] which emphasise that organisations are increasingly expected to contribute not only to economic performance, but also to environmental and social sustainability.

This development naturally follows from the core ideas introduced earlier in this chapter. If quality is defined by the customer, then organisations must consider not only immediate satisfaction but also long-term trust, reliability, and responsibility. If quality problems are systemic, improvements must address the underlying processes rather than short-term symptoms. If quality improvement requires deliberate planning, control, and structured change, then organisations must be capable of managing such changes over time.

When organisations consistently fail to meet expectations, waste and unnecessary costs often follow through rework, inefficiency, or lost value. Reducing such waste is therefore not only an economic priority, but increasingly also a societal concern in a world shaped by resource constraints, sustainability challenges, and long-term risk.

From this perspective, sustainability is not an external add-on to quality management, but an extension of its underlying logic. Reducing waste, improving resource efficiency, stabilising processes, and learning systematically from data all contribute simultaneously to lower cost and reduced environmental and social impact. Similarly, long-term thinking, stakeholder orientation, and risk awareness are consistent with the quality management tradition's emphasis on prevention, learning, and continuous improvement.

As organisations have become more interconnected and exposed to global competition, regulatory expectations, and societal scrutiny, quality management has been increasingly integrated with other management domains. As the scope of quality management has expanded, so too has its societal context. Quality questions are increasingly intertwined with broader concerns, such as environmental impact, resource use, social responsibility, and long-term risk. Issues such as food waste, overproduction, excessive consumption, and climate change highlight the limits of narrowly defined efficiency and short-term optimisation.

From a quality management perspective, these challenges raise familiar questions. How is value defined and for whom? What forms of waste are embedded in the current systems? How do organisations learn, adapt, and manage risk over longer time horizons? Value increasingly includes long-term societal impact and resource stewardship, in addition to immediate customer satisfaction. Addressing these questions requires the broader and more thoughtful application of quality management principles.

This perspective reinforces the importance of a quality culture. Sustainable performance depends on shared values, norms, and behaviours that encourage people to identify problems, learn from their experiences, and contribute to improvement. Sustainability highlights what quality management has long implied: lasting results require commitment, competence, and collaboration across the organisation.

Sustainability-integrated management represents a continuation of the roots of quality management. It reflects the same fundamental concerns with variation, value, cost, learning, and improvement, applied to a wider set of challenges and a longer time horizon.

Case: Competing on price, quality, and responsibility—A supermarket chain at a crossroads

A regional supermarket chain competes aggressively on price. Its strategic position in the market is built on offering customers consistently low prices, reliable availability of everyday products, and convenient locations to shop. Over the years, the company has invested heavily in logistics efficiency, standardised store layouts, and tight control of operating costs.

Recently, management has become increasingly concerned about the amount of food waste generated in stores, particularly fresh produce such as fruits and vegetables. Large quantities of food are discarded every day due to over-ordering, short shelf life, and fluctuating customer demand. While waste levels have long been accepted as a normal cost of doing business, rising disposal costs, increased public attention to sustainability, and internal pressure from employees have made waste management more visible.

The company monitors waste levels through routine reporting. Store managers are expected to stay within predefined limits, and deviations are discussed in regular performance meetings. Occasionally, sharp increases in waste can be traced to specific events, such as delivery errors, refrigeration failures, or promotional campaigns that did not generate expected sales. These incidents are handled locally, and corrective actions are taken.

However, even when no abnormal events occur, the overall waste levels remain high. Senior management now recognises that the problem is not simply one of control, but of how the underlying processes are designed. Forecasting methods, ordering routines, supplier agreements, pricing strategies, and in-store handling practices all contribute to the outcome. Reducing waste requires changes across functions and over time.

A proposal has been put forward to launch a structured improvement initiative focused on reducing food waste while maintaining low prices and ensuring customer satisfaction. The improvement team would consist of representatives from procurement, logistics, store operations, marketing, and IT. This effort would require data analysis, testing of alternative solutions, and coordination across organisational boundaries.

Management is divided. Some argue that waste is inevitable in low-price, high-volume businesses and that further efforts would increase costs. Others believe that reducing waste is both an economic opportunity and a responsibility to customers, employees, and society. The leadership team must decide how to proceed.

Reflection points—Chapter 2

1. Standardisation and human judgement
   A growing organisation introduces detailed procedures and performance metrics to ensure consistent quality across multiple locations. Over time, employees begin to follow instructions mechanically, and their initiative declines. How should management balance the benefits of standardisation with the need for professional judgement and learning? Where is the boundary between necessary discipline and counterproductive rigidity?
2. Reacting to performance fluctuations
   A service provider experiences recurring fluctuations in delivery times and responds by frequently changing the staffing levels and issuing corrective memos. Despite constant adjustments, the overall performance remained unstable. How can leaders determine whether the problem lies in normal process variation or in specific underlying causes? What risks arise when management reacts to every deviation without understanding the system?
3. Control or improvement?
   A company closely monitors key performance indicators and responds quickly to deviations; however, its long-term performance does not improve. Managers spend most of their time "keeping things under control" rather than redesigning processes. How can leaders distinguish between issues that require tighter control and those that demand structured improvement? When does stability become stagnation?
4. Efficiency, sustainability, and long-term responsibility
   A firm reduces operational waste primarily to cut costs and later discovers that the initiative also improves environmental performance and public reputation. Should such actions be justified purely on economic grounds, or does quality management require a broader societal perspective? How can organisations reconcile short-term efficiency with long-term legitimacy and resilience?
5. Quality and sustainability trade-offs
   A company succeeds in reducing defect rates but achieves this through energy-intensive processes and higher material consumption. Has quality improved from a sustainability perspective? Discuss how expanding the definition of quality to include long-term resource use changes managerial decision-making.
6. Stability in turbulent environments
   An organisation operating in a volatile market implements structured procedures to stabilise operations, but employees complain that the system reduces flexibility and slows adaptation. Using the foundational logic of quality management introduced in this chapter, discuss how structure and adaptability can coexist. When does standardisation strengthen resilience, and when might it hinder it?

Notes

1 Taylor, F.W. (1911). *The Principles of Scientific Management*. New York: Harper & Brothers.
2 Shewhart, W.A. (1931). *Economic Control of Quality of Manufactured Product*. New York: D. Van Nostrand Company.

3 Deming, W.E. (2000). *Out of the Crisis*. Cambridge, MA: MIT Press.
4 Juran, J. (1989). *Juran on Leadership for Quality: An Executive Handbook*. New York: Free Press.
5 Feigenbaum, A.V. (1991). *Total Quality Control*, 3rd edition. New York: McGraw-Hill.
6 Crosby, P. (1980). *Quality is Free – The Art of Making Quality Certain*. New York: New American Library.
7 United Nations. (2015). Transforming our world: The 2030 agenda for sustainable development. https://sdgs.un.org/2030agenda.

3 Understanding systematic quality management

3.1 Why systematic quality management matters

Organisations are open systems that continuously interact with their environment. Quality management only makes full sense when it is approached systematically. Materials, information, people, expectations, regulations, and resources flow into the organisation, are transformed through various processes, and result in products or services that affect customers, suppliers, authorities, and society at large. Therefore, quality is not created at a single point, nor can it be secured by controlling one function in isolation.

If quality is influenced by how processes interact, how responsibilities are coordinated, and how expectations are interpreted across the organisation, then quality management must address the organisation as a whole. A fragmented or ad hoc approach may work for simple operations, but it quickly becomes insufficient as complexity increases. A systematic approach helps an organisation understand how different activities are connected, where interfaces exist, and how decisions in one part of the system affect outcomes elsewhere. What may follow is consistency in operations, learning from experience, and continuous improvement.

Systematic quality management is also closely linked to prevention. Many quality-related problems arise not because individuals perform poorly, but because expectations are unclear, responsibilities are poorly defined, or processes are not coordinated. By establishing a systematic framework for how work is planned, executed, and followed up, an organisation can reduce the likelihood of errors and misunderstandings before they occur. Systematic quality management is about designing the organisation in a way that makes problems less likely in the first place.

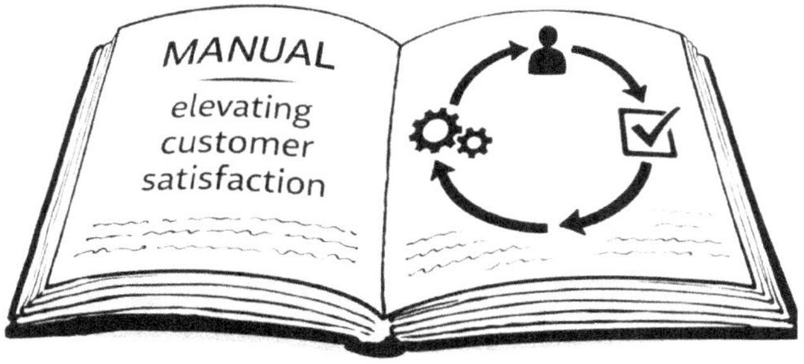

Figure 3.1 A documented QMS.

DOI: 10.4324/9781003671749-3

Finally, a systematic view is essential for understanding risks. Risks in organisations usually emerge at interfaces, in handovers, in dependencies between processes, or in mismatches between internal practices and external expectations. To identify, understand, and manage such risks, an organisation must first have an overview of how it operates as a system. A systems approach provides this overview and forms an important foundation for responsible and sustainable organisational performance.

From this perspective, quality management systems (QMSs) can be understood as a practical and concrete way of applying systems thinking to quality management.

3.2 Quality management systems and organisational practice

A management system can be defined as a set of policies and processes that enable an organisation to operate in a controlled and purposeful manner and achieve its objectives. In this book, the focus is more specific, namely, on how quality-related expectations are defined, communicated, fulfilled, and demonstrated. The terms *quality management system*, *QMS*, and *quality system* are used interchangeably.

A quality management system (QMS) can be understood as a management system used to direct and control an organisation with regard to quality (International Organization for Standardization [ISO], 2015).[1] In practice, it consists of a set of policies, processes, responsibilities, and procedures that guide how an organisation fulfils expectations and coordinates its activities. At the core of this definition lies a very simple idea. An organisation issues a promise regarding how its operations will take place and how its products or services will meet expectations. This promise may be directed at customers, but it also concerns owners, employees, authorities, and society at large.

Therefore, several interested parties have a direct interest in an organisation operating a QMS. For internal stakeholders—such as managers, owners, and employees—the system helps coordinate activities and clarify responsibilities. This can lead to better organisation of work, more consistent results, and improved communication across the

Figure 3.2 Creating a sense of assurance with simple documentation.

organisation. For employees, a QMS can contribute to a more transparent work environment and provide useful guidance when new staff are introduced to the organisation.

From the perspective of external stakeholders, such as customers, suppliers, and business partners, a QMS provides quality assurance. It creates confidence that the organisation can fulfil expectations consistently and in a controlled manner. This is particularly important when organisations operate in competitive markets or when customers are unfamiliar with the company's reputation. A visible and functioning QMS signals that the organisation has established methods for controlling its processes, documenting its activities, and demonstrating that requirements are fulfilled.

In practical terms, a QMS can contribute to operational efficiency. By clarifying how tasks are delegated, reducing misunderstandings, and ensuring that important aspects of the operation are not overlooked, the system helps to prevent errors and unnecessary rework. It also makes auditing possible and increases the transparency of the organisation's activities, which in turn strengthens trust among stakeholders.

From this perspective, it is difficult to find an organisation that does not, in practice, have some form of quality system. Any organisation that has been in operation for some time must be able to deliver what it promises to its customers. Customers continue to return, employees know how work is done, and managers can coordinate activities. This would not be possible without an underlying system that guides behaviour and decision-making. Such a system does not necessarily have to be formal, documented, or explicitly discussed. It may consist of established routines, shared understandings, and unwritten rules that have evolved and become embedded in the organisational culture.

In many small or stable organisations, informal systems can function surprisingly well. Employees know what is expected of them, tasks are carried out in a familiar way, and quality is maintained through experience and professional pride.

A simple example illustrates this point. One summer morning, the author of this book heard a knock on his door. Outside stood two young entrepreneurs: a 12-year-old boy and his 11-year-old sister, offering lawn-mowing services in the summer. They presented a one-page document describing who they were, what services they offered, how often the lawn would be mowed, what equipment would be used, the price, and how they could be contacted. They explained their proposal, asked the author to consider it, and promised to return the next week.

This was a very simple "organisation," and certainly not a formal one. Nevertheless, what the children presented was, in effect, an informal quality management system. They issued a clear promise, described how the work would be carried out, and demonstrated their credibility. This was sufficient to create confidence that expectations would be met. They got the job for the summer and delivered what they had promised.

This example supports the central idea that a typical organisation has a quality system, whether formal or informal. The difference lies not in whether a system exists, but in how explicit, structured, and robust it is. As organisations grow, operations become more complex, expectations multiply, and risks increase. Under such conditions, informal systems are often insufficient. This is where the need for more formalised quality management systems emerges.

3.3 From informal systems to formalised quality management

When organisations are small, operations are simple, and the same individuals are closely involved in most activities, a shared understanding and experience—or even standardised training of the staff—can be sufficient to ensure that expectations are met. Quality is maintained through professional judgement, direct communication, and familiarity with the work. In such settings, formal documentation may add little value and can be perceived as unnecessary bureaucracy.

This situation changes as organisations grow and their operations become more complex. As more people become involved, roles become specialised, tasks are delegated, and decisions are made at multiple levels. At the same time, expectations from customers, authorities, and society tend to become more explicit. Under these conditions, reliance on tacit knowledge and unwritten routines is increasingly risky. Ensuring consistency, maintaining clear communication across interfaces, and coping with staff turnover are significant challenges when expectations are no longer shared informally but must be understood and applied by many individuals.

As a result, organisations gradually move towards more formalised quality management, where expectations, responsibilities, and ways of working are made explicit. Documentation then becomes a supporting mechanism for coordination, continuity, and transparency, and a means to ensure that the organisation can operate reliably as a whole.

We can also consider environments where the consequences of failure are so severe that extensive formalisation is unavoidable. A typical example of this can be found in the aviation industry. The maintenance of jet engines in commercial aircraft must follow strictly defined procedures, and every action must be documented to ensure full traceability. It must be possible to demonstrate who performed the task, when it was performed, which components were used, and which requirements were applied. In such contexts, informal routines or undocumented decisions are unacceptable, as safety and public trust depend on the ability to provide clear proof that all requirements have been met.

These examples illustrate the wide range of quality management systems that are currently in practice. Most organisations operate between these extremes. When quality management is implemented at the organisational level, the whole organisation should move towards a more formalised system. Different parts of the organisation may require different levels of detail and documentation, depending on risk and external demands; however, these differences must exist within a coherent and shared quality management system.

Figure 3.3 Formalised quality management.

3.4 Promise, execution, and proof: The core logic of a QMS

At the heart of any quality management system lies a simple but powerful logic that can be described through three closely related elements: the promise, the execution of the promise, and the proof that the promise has been fulfilled. Together, these elements form the core of systematic quality management and provide a common reference point.

Promise refers to what the organisation commits to operate and deliver. It may be expressed in different forms and at different levels, ranging from general policy statements to more specific descriptions of products, services, processes and responsibilities. A promise creates expectations among customers and other stakeholders. Once such expectations are created, they should be met in a consistent and reliable manner.

Execution concerns how an organisation performs its work. Stating intentions or defining procedures is not enough; the work must be carried out accordingly. Employees need to understand what is expected of them, responsibilities must be clearly assigned, and the necessary resources and competencies must be available. Execution, therefore, links the abstract commitments of the organisation to everyday operational practice.

The third element, proof, distinguishes systematic quality management from good intentions alone. If an organisation claims that it operates in a certain way, it should also be able to demonstrate that this is indeed the case. Proof is created through records, data, and other forms of evidence generated by the operation of the system. Completed checklists, filled forms, reports, contracts, meeting minutes, and measurement results serve as evidence.

These three elements are inseparable. A promise without execution has no value, execution without a clear promise lacks direction, and both are weak if there is no proof that they have been executed. Together, they create transparency, make it possible to assess whether expectations are fulfilled, and support learning, auditing, and improvement. This triad also explains the role of documentation in quality management systems.

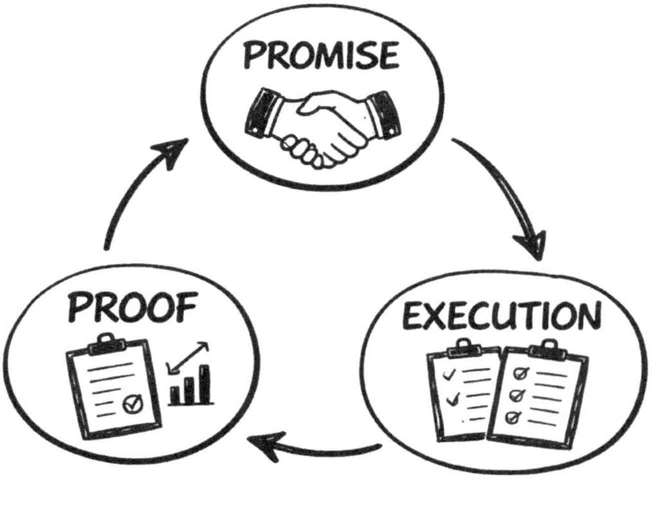

Figure 3.4 The heart of every QMS: promise, execution, and proof.

Documents are created to make promises explicit, support consistent execution, and provide evidence.

3.5 Categories of quality management systems

So far, quality management systems have been discussed in rather general terms. In practice, however, quality systems can take many different forms, depending on the size of the organisation, the complexity of its operations, the level of risk involved, and the expectations of interested parties. To clarify, it is useful to distinguish between five categories of quality management systems, arranged in a logical sequence that reflects increasing levels of formality:

- Informal procedures by the master. A professional takes responsibility for his or her own work for the customer, even if there is no formal description or quality manual.
- Registered policies and main procedures. The main procedures and policies are documented, meaning that a formal and explicit promise is issued to the customer.
- Formal registration of the data. Data are systematically registered in the operation of the QMS, for instance, confirmations that the properties of a product or service are in accordance with specifications.
- Internal audits. The operation of the QMS includes internal audits that assess whether the system is in use, whether it is run in an active way, and how relevant it is.
- External audit and certification. The QMS is audited by an external auditor with appropriate credentials, providing formal and independent confirmation that it fulfils the defined requirements.

These five categories build on each other. For example, systematic registration of data presupposes that documented procedures exist, and meaningful audits are only possible when there is a clear promise and evidence of their execution.

The focus of this book is on quality management systems in the fifth category, where formal requirements are defined, and compliance is verified through independent assessment. Nevertheless, the underlying logic of promise, execution, and proof applies to all categories.

3.6 QMS and the organisation as an open system: Internal and external forces

A quality management system can only be fully understood when the organisation is viewed as an open system, operating under the influence of various forces in its internal and external environment. Organisations do not operate in isolation; they are

Figure 3.5 Five categories of QMS

continuously affected by expectations, requirements, and constraints that originate both within and outside the organisation. Materials, information, people, and resources flow into the organisation, are transformed through its processes, and result in products or services that influence customers, suppliers, authorities, and society at large.

Within an organisation, several internal forces are at work. These forces are represented by different interested parties, each with its own expectations and demands. Owners or shareholders expect the organisation to fulfil its defined purpose, whether it involves profitability, public service, or other objectives. They also expect compliance with laws and regulations, reliable reporting, and a good reputation. Managers expect employees to act professionally, show integrity and initiative, and deliver results in line with organisational objectives. To do so, managers depend on clear information, coordination and predictable processes. Employees, in turn, expect support from management, clear direction, proper training, fair treatment, and a safe and healthy working environment.

Simultaneously, the organisation is subject to external forces arising from its environment. Customers expect products and services that meet the agreed requirements, are delivered on time, and comply with relevant standards and regulations. Suppliers expect clear specifications, timely communication, and feedback on performance. Beyond these direct relationships, there are broader forces in society, including authorities, institutions, unions, associations, and the general public. These reflect expectations regarding transparency, responsibility, safety, environmental performance, and ethical conduct. In today's business environment, where information spreads rapidly through traditional and social media, such external forces can have a strong and immediate impact on organisational operations.

A quality management system provides a structured approach to address and balance these internal and external forces. By defining policies, procedures, and responsibilities, it helps translate diverse and sometimes conflicting demands into coherent organisational practices. It makes explicit how expectations are interpreted, how work is organised, and how compliance is ensured.

The QMS functions as a system-level mechanism that links an organisation to its environment. It supports internal coordination, facilitates external communication, and provides a transparent basis for demonstrating that the organisation operates in a controlled and responsible manner. Understanding these forces and their interactions is a key prerequisite for systematic quality management.

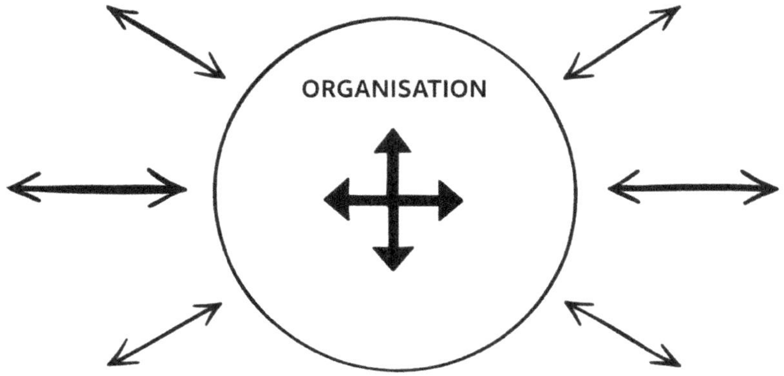

Figure 3.6 Forces at work in the internal and external environment.

3.7 Understanding purpose, expectations, and risk through a QMS

One of the most important contributions of a QMS is that it helps an organisation understand and manage its purpose and expectations in a structured manner. Organisations exist to fulfil a defined purpose, which is translated into expectations by various interested parties. These expectations may relate to quality, usability, safety, environmental performance, reliability, cost, or ethical conduct. A QMS provides a framework for identifying these expectations, interpreting them consistently, and embedding them in the organisation's way of working.

The likelihood of problems increases when expectations are unclear, contradictory, or poorly communicated. Products may fail to meet customer needs, internal conflicts may arise, and legal and regulatory requirements may be overlooked. From this perspective, many quality-related problems can be understood as manifestations of risk—the risk that the organisation will fail to fulfil its promises or meet the expectations placed upon it.

Risk in organisations typically emerges at interfaces between departments, between the organisation and its suppliers, between management intentions and operational practices, or between internal routines and external requirements. By creating an overview of how an organisation operates as a system, such risks are more easily identified and managed. A holistic view is crucial; local optimisation in one part of the organisation may increase risk elsewhere if interactions and dependencies are not understood.

A quality management system supports risk awareness by providing a coherent and shared description of how an organisation functions. By making promises explicit, clarifying responsibilities, and defining processes, it reduces ambiguity and creates a common understanding of what is expected and how it should be achieved. The systematic use of procedures, records, and reviews also makes it possible to detect deviations and emerging problems before they develop into serious failures.

Thus, a QMS can be considered a preventive mechanism. It does not eliminate uncertainty or guarantee perfect outcomes, but it reduces the likelihood of undesirable events by addressing their underlying causes. This is directly linked to the earlier discussion on prevention and quality costs. Investing effort in systematic quality management is often far less costly than dealing with the consequences of failures after they occur. By providing an organisational overview and supporting informed decision-making, the QMS becomes an important tool for responsible and sustainable management.

3.8 The structure of a QMS: The quality pyramid

The structure of a QMS can be illustrated simply and intuitively using a quality pyramid. This simple model helps explain how different elements of the system relate to one another and how they support the fulfilment of expectations. The pyramid is divided into four levels, each representing a different type of content in the QMS.

Policy documents are at the top of the pyramid. These describe what the organisation does, what it stands for, and where it is heading. The quality policy, in particular, expresses the organisation's overall commitment to quality and outlines the principles that guide its operation. At this level, the organisation defines its intentions and makes general promises to customers and other interested parties.

The next level of the pyramid consists of the work procedures. While policies state what the organisation commits to, procedures explain who is responsible for doing what, where activities take place, and when they are to be performed. These procedures are

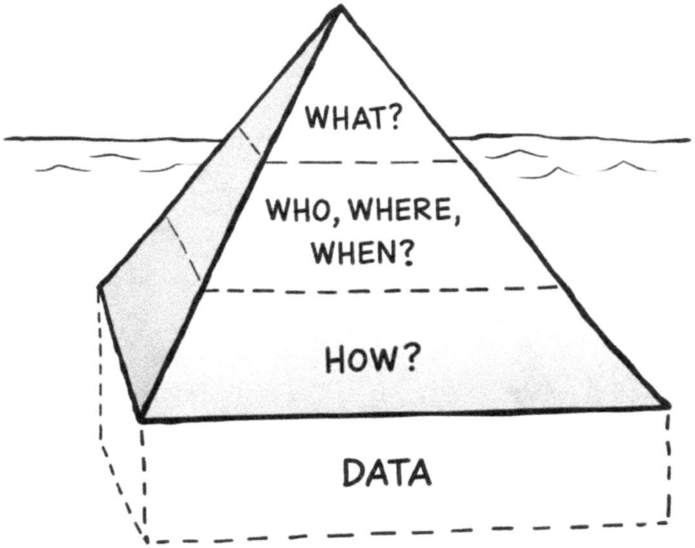

Figure 3.7 The quality pyramid.

typically general in nature and apply across the organisation. They translate policy commitments into organised and repeatable ways of working. Work procedures can be presented in various forms, such as flowcharts, tables, or textual descriptions; however, they should always be clear, simple, and accessible to those who are expected to follow them.

The third level addresses the question of how the work is carried out in practice. In many cases, general procedures are insufficient as direct guidance for daily tasks. More detailed instructions, checklists, technical guidelines, or forms may be required to support employees in specific situations. These can be referred to as support documents, and they are closely linked to the relevant work procedures. Together with policies and work procedures, they form what we call the quality documents of the system.

At the base of the pyramid lies the evidence generated through the operation of a quality management system. This includes completed forms, checklists, reports, contracts, meeting minutes, measurement results, and other records created as work is performed. These records provide proof that the organisation has operated in accordance with its stated policies and procedures. They are not created for documentation's sake but as a natural outcome of systematic execution.

An important and somewhat paradoxical aspect of the quality pyramid is that, unlike a physical structure, it must be built from the top down. An organisation must first define its policies and promises before it can meaningfully design procedures. Only after procedures are established does it make sense to develop supporting documents. The foundation of records and data then emerges naturally as the system is used in practice. Starting with forms and records without procedures and clear policies would lead to confusion and inefficiency rather than effective quality management.

3.9 QMS as a living handbook for the organisation

A quality management system should never be understood as a static collection of documents or an administrative burden imposed on the organisation. At its core, a QMS is a

Figure 3.8 QMS as a living document.

handbook for employees—a shared reference that explains how the organisation works, what is expected, and how tasks should be carried out in practice. For this reason, the system must be designed with users in mind and developed with their participation.

Employee involvement is essential. Those who perform work on a daily basis possess critical knowledge about processes, interfaces, and practical constraints. Without their input, quality documents risk becoming detached from reality and are difficult to apply. Participation also creates ownership, which increases the likelihood that the system is understood, trusted, and used in everyday work.

Equally important is the recognition that the QMS is a living system. Organisations operate in environments that change over time: customer expectations evolve, regulations are updated, technologies develop, and business models are reshaped. Therefore, the QMS must be continuously reviewed and revised to remain relevant and useful.

A simple example illustrates this point. Consider an IT company that provides services to public institutions. Such organisations operate under strict and evolving requirements related to data security, privacy, and protection of personal information. Increased regulatory attention to issues such as information security and data protection may require new procedures, revised responsibilities, and additional controls. At the same time, rapid technological development can fundamentally change how services are delivered, alter risk profiles, or even transform the competitive landscape. If the QMS is not updated to reflect these changes, it quickly becomes outdated and may hinder effective operations. However, by providing both structure and adaptability, the QMS enables the organisation to remain stable in changing conditions, maintaining reliability while adjusting to new expectations and risks.

This example illustrates why a QMS cannot be treated as a fixed set of documents. It must evolve in response to changes in the organisation's environment, while at the same time providing stability and clarity in daily operations. This dynamic character links the QMS directly to earlier discussions on open systems, expectations, and improvement.

Finally, it is important to emphasise that the objective of a QMS is not to produce extensive documentation. Only those documents that are necessary to ensure that expectations are met consistently should be included. If a document does not add value or is no longer relevant, it should be revised or removed. The form and format of the QMS must also be adapted to the organisation's culture and way of working, whether this involves traditional manuals or digital platforms that allow easy access and continuous updating of the QMS.

A small observation from practice illustrates this point. During a strategic planning discussion in a construction company, the results of interviews with members of the management team were summarised. The company operates with a certified quality management system and complies with several international ISO standards. Two relatively new employees were present mainly to listen and learn. At one point, they were asked whether the QMS had helped them understand how the organisation operates or how their work should be carried out. After a brief pause, they admitted that they had not used the system.

Such situations do not necessarily indicate that the quality management system was poorly designed. In many organisations, systems have been developed in good faith and in accordance with recognised standards. A more important question is whether the system remains relevant, useful, and connected to actual practice. If employees do not turn to the QMS when trying to understand how the organisation works, this may indicate that the system has gradually become detached from what is perceived as normal practice.

A QMS needs to be a living handbook, an integral and evolving part of how the organisation functions, supporting coordination, learning, and improvement while simultaneously providing stability and clarity in daily operations. In many organisations, this evolution increasingly takes place in digital environments, where documentation, coordination, and monitoring are integrated through interconnected information systems.

Case: When a quality management system becomes a formality

A regional construction contractor implemented a certified QMS after public clients began requiring certification as a condition for participating in the tenders. External consultants supported the process, documentation was prepared, procedures were written, and certification was obtained easily. The company has since maintained its certification through regular internal and external audits.

The organisation also takes pride in being strongly safety oriented. Worker safety is deeply embedded in daily routines and professional conduct in the construction industry. Interestingly, much of this commitment is rooted in culture and leadership rather than in formal quality management systems.

On paper, the QMS is complete and well-structured. In daily practice, project managers and site supervisors primarily rely on their experience and established routines. Some descriptions in the system no longer reflect current methods, and several templates are seldom used. New employees typically learn "how things are done" through colleagues who have been in the organisation for some time rather than through documented procedures.

Internal audits are carried out as required and generally confirm that the formal system is in place. Deviations are usually minor and are corrected before external audits. Although certification is maintained, the QMS plays a limited role in guiding everyday decisions or supporting coordination across projects.

This misalignment became visible when recurring issues appeared in recent projects, such as inconsistent subcontractor follow-up and incomplete handover documentation. None of these represent major failures, but together, they suggest that expectations are not always translated into consistent practice. Senior management has begun to reflect on the role of the QMS: Is it primarily a mechanism for maintaining certification, or should it function as a shared operational reference for the organisation?

The company is now at a crossroads. It can continue to maintain a formally compliant system that fulfils external requirements, or it can rethink its QMS as a living handbook—one that clarifies expectations, strengthens coordination, and supports organisational learning.

The question is not whether a company has a quality management system. It clearly does. The more fundamental question is what purpose the system is intended to serve and whether it remains a meaningful part of how the organisation operates.

Reflection points—Chapter 3

1. Compliance or operational value
   A mid-sized, project-oriented company has maintained ISO certification for several years, yet its managers increasingly rely on informal routines rather than documented procedures. External audits showed no major deviations; however, recurring coordination problems persist across projects. As a senior management member, how would you assess whether the QMS primarily serves compliance or operational value? What risks arise if the formal system and daily practice continue to diverge, and how might this affect the long-term performance?
2. From promise to proof
   A public agency publishes an ambitious quality policy promising transparency, reliability, and user orientation, but its internal procedures are loosely defined, and performance data are rarely reviewed. How would you analyse this situation using the logic of a promise, execution, and proof? Where are the most critical weaknesses likely to occur, and what would be the consequences for legitimacy and trust if the proof does not match the promise?
3. Internal and external forces in tension
   Manufacturing companies face pressure from owners to reduce costs, employees to improve working conditions, and customers to enhance product reliability. Using the open-system perspective, how would you analyse these internal and external forces? How can a QMS help balance competing expectations, and what risks emerge if one set of forces dominates the other?
4. A living system in a changing environment
   An IT service company serving public institutions needs to adapt to rapidly changing data protection regulations and technological developments. Although daily practices have evolved informally, its documented procedures have not been updated for several years. What are the potential consequences of this gap between formal documentation and actual operations? What must be considered when redesigning the QMS?
5. Documentation—necessary or excessive?
   A newly appointed quality manager discovers that the organisation's QMS contains hundreds of documents, many of which are rarely used. Removing them may increase clarity but could also create uncertainty regarding compliance. How would you decide which

documents are truly necessary? What criteria would you use to evaluate whether the documentation supports prevention and coordination rather than merely satisfying external audits?

6. Risk at the interfaces
   Several minor failures in a construction project are traced not to technical errors but to misunderstandings between departments and subcontractors. How would you analyse these issues from a system-level perspective? In what ways could a well-designed QMS reduce risks emerging at organisational interfaces, and what unintended consequences might arise if controls become overly rigid?

Note

1 International Organization for Standardization. (2015). ISO 9000:2015 Quality management systems—Fundamentals and vocabulary (ISO Standard No. 9000:2015).

4 Applying international standards and frameworks

4.1 Standards as a practical reference when building a QMS

When organisations seek to improve their operations and manage quality systematically, they are often confronted with a wide range of choices. One such choice is the use of an international standard or a recognised framework as a reference. The use of a standard is not a prerequisite for quality management or a necessary condition for building an effective QMS. Many organisations operate successfully without formal reference to any management standards. At a fundamental level, a standard can be understood as an agreement among different stakeholders regarding what constitutes an appropriate or generally acceptable solution, in a given context. Therefore, standards can be useful for different reasons.

One good reason why international management standards may be helpful is that they represent a consolidated view of what is generally regarded as sound practice in organising, managing, and controlling processes, products, or services so that they deliver what is expected and are fit for purpose. And it should be noted that the development and maintenance of international management standards such as ISO 9001 (ISO, 2015)[1] represent an extraordinary collective effort. Each major revision typically involves several hundred experts from dozens of countries, working over several years and reviewing thousands of comments through formal consensus processes.

Another important reason for using standards relates to quality assurance and trust. From the perspective of customers and other interested parties, the application of a recognised standard can signal that an organisation manages its operations in a controlled and systematic manner. This assurance is sometimes explicitly demanded by clients, regulators, or partners. In other cases, it may simply strengthen confidence in the organisation and reduce uncertainty in transactions. The use of recognised standards may also strengthen organisational legitimacy by signalling responsible and structured management not only to customers, but also to regulators, partners, and society more broadly.

It is also important to distinguish between the two fundamental types of standards. Some are designed for certification and define explicit requirements that independent auditors can assess objectively. The Others are guidance standards that provide recommendations and examples of good practices without offering certification.

When an organisation adopts standards for the wrong reasons or interprets requirements as rigid prescriptions rather than adaptable guidance, a quality management system may become burdensome and disconnected from everyday practice. In such cases, the system risks serving compliance rather than supporting the operational value. This underlines the importance of treating standards as references to be interpreted considering the organisation's purpose, context, and manner of working.

DOI: 10.4324/9781003671749-4

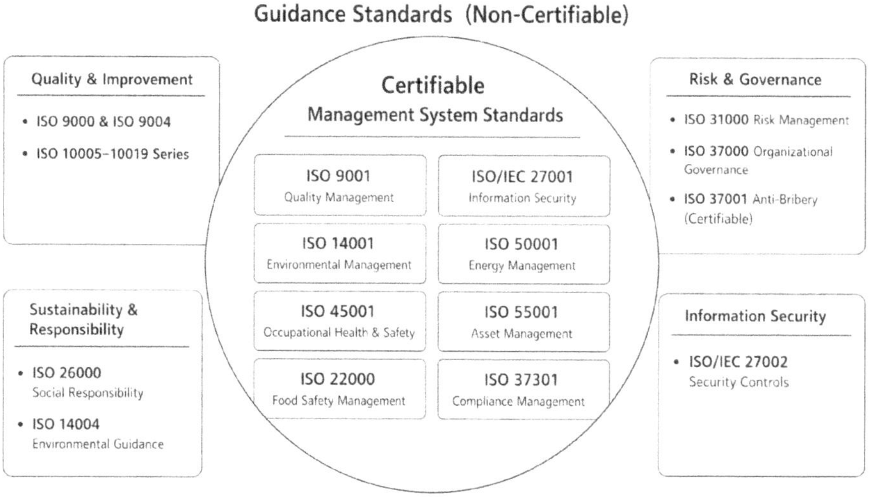

Figure 4.1 Certifiable ISO management system standards and ISO guidance standards.

In some situations, it may be more appropriate to use a quality management framework rather than a certifiable standard. Frameworks do not lead to certification, but they emphasise reflection, learning, and organisational development. They can also provide a shared language for discussing performance and improvements. In practice, these approaches are often combined. Standards and frameworks should therefore be seen as complementary tools, whose value depends on how they are interpreted, adapted, and applied in the context.

In this chapter, ISO 9001 is used as the primary reference for explaining how international management standards can be applied to build a management system. This choice is not meant to suggest that quality management is limited to this standard. Rather, ISO 9001 provides a clear and widely recognised example of a systematic approach to management. Building on this foundation, the chapter also introduces related standards, such as ISO 14001 and ISO 45001, as well as broader quality management frameworks that organisations can use as references when developing systematic approaches to quality management.

The chapter revisits ideas introduced earlier in the book, such as the PDCA logic and fundamental quality management principles, which are central to ISO 9001. Other key elements of the standard—such as internal audits and management review—are introduced here and explored in more detail in later chapters.

4.2 ISO and the evolution of the ISO 9001 standard

International management standards are developed within the framework of the International Organization for Standardization (ISO), a global network of national standards bodies. ISO management standards are living documents that are periodically revised to reflect changes in knowledge, technology, and expectations in relation to how organisations are managed. As a result, they are applied by a wide range of organisations, including production companies, service providers, public institutions, and non-profit organisations.

The ISO 9001 standard belongs to the ISO 9000 family of standards that focuses on quality management. Its origins can be traced back to the British Standard BS 5750, published in 1979, which provided a model for quality assurance in manufacturing. When the first version of ISO 9001 was issued in 1987, it largely reflected this background and was primarily oriented towards production environments. Subsequent revisions gradually broadened the scope of the standard and extended its applicability to other types of organisations.

A significant step in this evolution occurred with the revision published in 2000, marking the transition from assurance-oriented compliance to a broader management perspective. The process approach was established as a central organising principle, and continual improvement became a core theme. The standard shifted its emphasis towards understanding how activities are interconnected, how value is created through processes, and how performance can be improved over time. Subsequent revisions refined this logic, strengthened the consistency of terminology, and reinforced the role of management responsibility and system coherence.

The revision published in the mid-2010s placed greater emphasis on the context of the organisation, including internal and external factors that influence its ability to achieve intended outcomes. Risk-based thinking was introduced as an explicit concept, reflecting a broader understanding of management as an ongoing balancing act between objectives, uncertainties, and opportunities rather than a search for complete control.

The current form of ISO 9001 builds on this long development path. The standard is applicable to all types of organisations, regardless of size, sector, or ownership. Its requirements are formulated at a high level, allowing organisations to interpret and apply them in ways that fit their specific context. Understanding this evolution helps explain why ISO 9001 is often regarded as a useful reference for building a quality management system today.

ISO 9001 is firmly grounded in traditional quality management ideology and the broader understanding of quality management introduced in earlier chapters of this book. In ISO terminology, these ideas are formalised as quality management principles, which underpin the requirements of the standard.

Figure 4.2 Leadership, customer focus, and engagement of people are at the forefront of ISO 9001.

Customer focus

At the core of these principles is a strong focus on customer and other interested parties. Organisations exist to fulfil a purpose, which is expressed through expectations regarding products, services, and behaviour. ISO 9001 emphasises the need to understand current and future customer requirements, as well as the relevant expectations of other interested parties, to meet these requirements consistently and to monitor customer satisfaction.

Leadership

ISO 9001 explicitly places the responsibility for the quality management system with top management. Leaders are expected to establish unity of purpose and direction, define objectives, and create conditions under which people can contribute effectively. Without active leadership and commitment at the highest level, quality initiatives tend to become fragmented, symbolic or short-lived. This leadership responsibility is closely linked to the development of a quality culture and ethical behaviour in how the organisation operates and makes decisions.

Engagement of people

Quality is not created by systems and documents alone, but through the actions and decisions of individuals throughout the organisation. ISO 9001 therefore emphasises competence, awareness and involvement. Employees at all levels are expected to understand their roles in achieving quality objectives and take responsibility for their contributions.

Process approach

The process approach provides a structural backbone of ISO 9001. The standard encourages organisations to understand their activities as interconnected processes that together form a coherent system for creating and delivering value. By identifying processes, clarifying their interactions, and managing them deliberately, organisations can achieve more predictable and consistent results.

Continual improvement

ISO 9001 assumes that no organisation operates in a stable environment and that performance must be reviewed and improved over time. Improvement is not limited to correcting errors or addressing nonconformities, but also includes learning, innovation, and adaptation. This principle reinforces the idea that quality management is an ongoing effort rather than a one-off implementation exercise.

Evidence-based decision-making

Decisions regarding processes, resources, and improvements are expected to be based on relevant data and information rather than intuition or assumptions. Therefore, monitoring, measurement, and analysis are integral to QMS. This principle supports transparency, accountability, and organisational learning.

Relationship management

Organisations depend on external parties, such as suppliers, partners, and service providers. Value creation often takes place across organisational boundaries, and effective quality management requires systematic communication and collaboration with these parties. This perspective reinforces the value chain logic and shows how quality management extends beyond the formal boundaries of an organisation.

The ISO 9000 family consists of four closely related standards.

- ISO 9000[2] provides the fundamental concepts, principles, and terminology used throughout the ISO 9000 family.
- ISO 9001[3] specifies the requirements for a quality management system and is the standard used for certification.
- ISO 9004[4] offers guidance for organisations that wish to go beyond basic requirements and focus on long-term success and maturity.
- ISO 19011[5] provides guidance on auditing management systems, including principles of auditing and audit programme management.

Together, these standards provide a coherent set of references that support both the implementation and ongoing development of systematic quality management.

4.3 ISO 9001: System model and the PDCA logic

ISO 9001 is a generic management standard. Its requirements are formulated at a high level so that they can be applied to organisations of different sizes, sectors and ownership forms. The standard does not prescribe in detail how an organisation must operate, nor does it define what constitutes a "good" product or service in any specific industry. Instead, it requires organisations to define their own objectives, understand their context

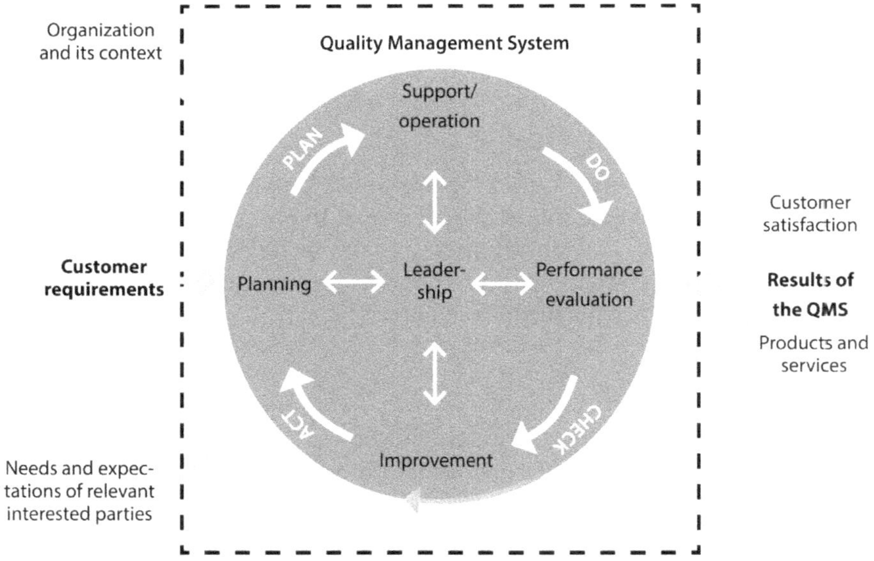

Figure 4.3 System representation of the ISO 9001 standard.

and the expectations of relevant interested parties, and establish a management system capable of delivering the intended outcomes.

A simple example helps clarify this distinction. McDonald's operates internationally, and many parts of its global organisation are certified according to ISO 9001. At the same time, there are mixed views among customers regarding its products. Some value consistency, predictability, and speed of service, while others criticise aspects such as nutritional content or perceived health impact. ISO 9001 certification does not resolve this debate, nor is it intended to do so. What the standard assures is that the organisation has a management system in place that enables it to define requirements, operate processes consistently, monitor performance, and improve systematically in line with its stated objectives.

This logic is best understood through the system model that underpins ISO 9001, which is built around the plan–do–check–act (PDCA) cycle.

The system model begins with an emphasis on the organisational purpose and context. Organisations are expected to understand why they exist, what they aim to achieve, and which internal and external factors influence their ability to deliver results. This includes identifying the needs and expectations of relevant interested parties, with particular emphasis on customers.

Based on this understanding, the organisation plans its quality management system. Planning involves defining the quality policy and objectives, determining the scope of the QMS, and establishing the processes, responsibilities, and controls necessary to achieve the intended outcomes. Planning also includes identifying risks and opportunities that may affect the organisation's ability to meet requirements.

These elements correspond to the phases of the PDCA cycle. Planning establishes direction and structure, doing represents the execution of operations, checking involves evaluating performance, and acting focuses on improvement. The "do" phase of the PDCA cycle corresponds to the everyday operations. Value-adding activities are carried out in accordance with defined processes and requirements, supported by the necessary resources, competence, infrastructure, communication, and documented information. The standard recognises that effective operations depend not only on core activities, but also on supporting functions that enable consistency, reliability, and control. These activities are expected to be managed deliberately, ensuring that the planned arrangements are implemented in practice.

The "check" phase addresses the performance evaluation. ISO 9001 requires organisations to monitor, measure, analyse, and evaluate their processes and outputs. This includes assessing customer satisfaction, reviewing process performance, conducting internal audits, and analysing data generated by the system. The purpose is learning, to understand whether the system is functioning as intended and where improvement is needed.

The "act" phase represents continual improvement, which is a central theme. Organisations are expected to respond to deviations, nonconformities, and changing conditions by correcting problems, addressing root causes, and improving the effectiveness of the management system. In practice, many improvement activities take the form of projects, linking quality management closely to project-based work.

Leadership plays a central role throughout this system model. Top management is responsible for maintaining the integrity of the QMS, ensuring alignment with the organisational strategy, and promoting conditions that support quality and learning. Through activities such as management review and engagement in improvement efforts, leadership ensures that the PDCA cycle remains active and that the system evolves over time rather than becoming a static set of procedures.

The ISO 9001 system model provides a coherent and flexible way to understand how organisations can manage quality systematically. It integrates purpose, planning, operation, evaluation, and improvement into a single framework aligned closely with the fundamentals of quality management principles. This perspective helps demystify the standard and clarifies why ISO 9001 serves as a useful reference for organisations seeking to build or strengthen a quality management system.

4.4 The aligned system framework

Over time, ISO has deliberately developed a common structure and shared logic for a family of management system standards. This structure is defined in Annex SL, which provides the mandatory framework for ISO management system standards such as ISO 9001[6] (quality management), ISO 14001[7] (environmental management), ISO 45001[8] (occupational health and safety), ISO 27001[9] (information security), ISO 22000[10] (food safety) and ISO 50001[11] (energy management).

The purpose of Annex SL is to ensure consistency in structure, terminology, and management logic across different domains. This reflects the recognition that organisations manage quality, environment, safety, and other concerns as interconnected aspects of a single system rather than as separate and unrelated activities. This alignment reinforces the view of an organisation as an open system embedded in regulatory, environmental, technological, and societal ecosystems.

All ISO management system standards that follow Annex SL share the same ten-clause structure. The first three clauses—scope, normative references, and terms and definitions—are all introductory. Clauses 4 to 10 contain the requirements and follow the plan–do–check–act (PDCA) logic introduced in the previous section.

- Clause 4—Context of the organisation. The organisation is required to understand its internal and external context, identify relevant interested parties and their needs, and define the scope of the management system. This ensures that the system is tailored to the organisation's environment and purpose.
- Clause 5—Leadership. Top management must demonstrate leadership and commitment, establish policies, assign responsibilities, and ensure accountability. Although the specific focus differs among the standards, the fundamental leadership role is consistent.
- Clause 6—Planning. Organisations must identify risks and opportunities, define objectives, and plan actions to achieve the intended outcomes. Risk-based thinking is a core element across all aligned standards and is embedded in planning rather than being treated as a separate activity.
- Clause 7—Support. This clause addresses the resources required to operate the management system, including competence, awareness, communication, and documented information. It provides a foundation that enables planned activities to be carried out consistently.
- Clause 8—Operation. The operational planning and control are addressed here. The specific content depends on the standard—for example, production and service provision in ISO 9001, environmental controls in ISO 14001, or emergency preparedness in ISO 45001—but the underlying management logic remains the same.

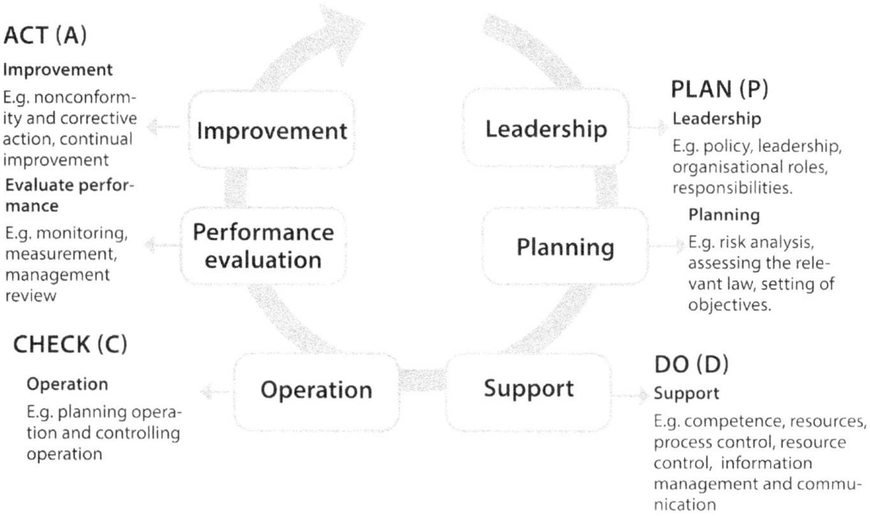

Figure 4.4 The common PDCA logic underlying ISO management system standards.

- Clause 9—Performance evaluation. Organisations are required to monitor, measure, analyse, and evaluate performance. This includes internal audits and management review, providing structured feedback on whether the management system is effective and fit for purpose.
- Clause 10—Improvement. This clause addresses nonconformities, corrective actions, and continual improvement. This ensures that the management system evolves over time based on experience, feedback, and changing conditions.

While the specific content varies between standards, the underlying structure and logic remain the same. This makes it easier for organisations to apply multiple standards in a consistent way and to integrate them into a coherent management system. Experience gained from implementing one standard—most commonly ISO 9001—is therefore largely transferable to others.

The following sections, 4.5 and 4.6, build on this aligned framework by exploring the common PDCA logic in more detail. ISO 9001 is used as the primary reference, providing a practical example of how the shared structure can be interpreted and applied in everyday organisational practice.

4.5 Planning and operation

Once the purpose, context, and overall logic of the quality management system have been established, ISO 9001 turns to how the organisation plans and operates its activities in practice. The focus is on translating intentions into action and organising everyday operations so that requirements are consistently fulfilled. In doing so, organisations must also consider external conditions that may influence their ability to achieve intended outcomes, including longer-term challenges such as climate-related impacts on operations, resources, and stakeholder expectations. This highlights how quality management increasingly extends beyond immediate customer requirements to broader sustainability considerations.

A central idea in this context is the distinction between value-adding and supporting activities. Value-adding activities are the processes through which the organisation creates its products or delivers its services to customers. They are directly linked to the organisation's purpose and customer expectations. Supporting activities do not create value directly from the customer's perspective, but are essential for enabling value-adding processes to function effectively, reliably, and consistently.

Value is created through a sequence of interconnected activities, and weaknesses in any part of this sequence can undermine the final outcome. ISO 9001 therefore requires organisations to identify, manage, and control both value-adding and supporting processes, and to understand how they interact as part of a coherent system.

A simple example illustrates this. In universities, teaching and research are the primary value-adding activities. These processes create value for students, society, and other stakeholders. However, they depend on a wide range of supporting activities such as administration, finance, IT services, libraries, facility management, human resources, and quality assurance. While these functions do not deliver educational or research outputs, their effectiveness strongly influences the quality, reliability, and sustainability of the university's core activities.

Planning the operation

Planning in ISO 9001 concerns how an organisation prepares its operations to deliver the intended results. This includes defining processes, establishing criteria for their operation and control, allocating responsibilities, and ensuring that the necessary resources are available. It also includes consideration of risks and opportunities that may affect the organisation's ability to meet requirements and integrate these considerations into operational arrangements.

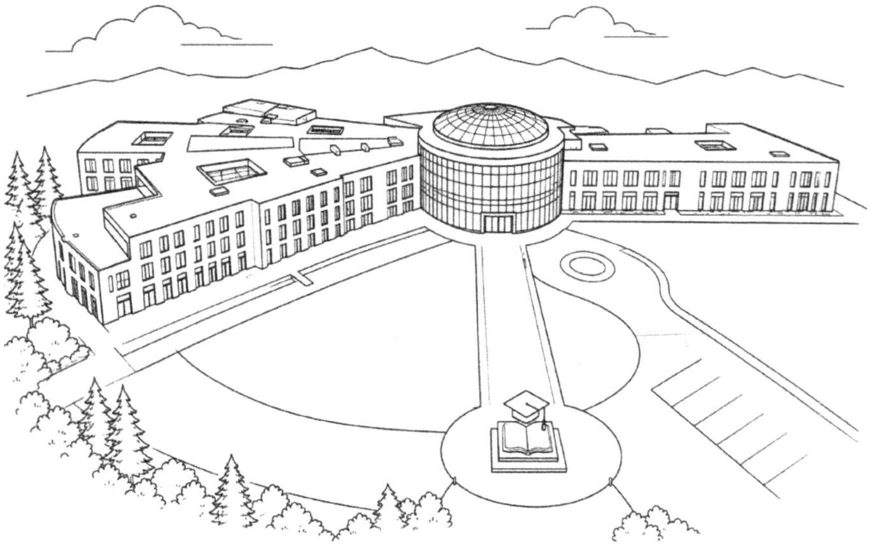

Figure 4.5 A university; teaching and research are primary activities.

Human resources are particularly important in this context. Organisations must determine the competence required for personnel performing work that affects quality, ensure that individuals are appropriately qualified or trained, and maintain competence records. Beyond formal qualifications, employees must also be aware of the quality policy, the objectives of the system, and how their work contributes to meeting requirements. Consequently, the standard emphasises awareness and communication across the organisation, recognising that quality is influenced not only by technical expertise but also by how people coordinate their work and interact with customers and stakeholders.

Consider a tourism company that offers specialised nature excursions. The competence of guides, their ability to communicate knowledge clearly, and their understanding of safety and environmental responsibility are all essential for delivering the promised service and maintaining customer trust in ecotourism.

In addition to human competence, organisations must ensure that the necessary infrastructure and working conditions are in place. Depending on the nature of the operation, this may include buildings, equipment, information systems, or specialised facilities that enable processes to operate reliably. The specific requirements differ across sectors, but the underlying logic is the same: the organisation must ensure that people, tools, and working conditions support the consistent delivery of products and services.

For example, a small earthworks contractor working on public infrastructure projects may require powerful machinery, reliable communication equipment for remote sites, and adequate facilities for workers at the construction site. In such a company, the behaviour and competence of employees on-site directly influence how the organisation meets technical requirements, safety expectations, and its public reputation.

Customer focus and operational control

ISO 9001 places customer requirements at the centre of the system. Organisations are required to determine customer requirements, including applicable legal and regulatory obligations, and to review them before committing to delivery. This review ensures that the requirements are clear, feasible, and aligned with the organisation's capability.

Where design and development activities are part of the operation, the standard requires these processes to be planned and controlled accordingly. Not all organisations have such activities in place. For example, an aluminium plant that melts scrap aluminium and casts standardised sows (a trapezoidal or rectangular, 500–2000 kg, moulded form of the metal) produces a commodity product with predefined characteristics; no design or development is involved. In contrast, an aluminium smelter that adjusts the alloy composition and casting parameters to meet different customer requirements must design and develop its products and processes. ISO 9001 therefore requires design and development controls only where the organisation is responsible for defining product or process characteristics.

Production and service provision must be carried out under controlled conditions. This includes the use of appropriate instructions, competent personnel, suitable equipment, and monitoring or measurements where necessary. Nonconforming outputs must be identified and controlled to prevent their unintended use or delivery.

Where monitoring or measurement is required, organisations must ensure that the equipment used provides reliable results. For example, an electricity transmission company must verify that its measuring instruments accurately record variables such as frequency, voltage, and power, because these measurements determine whether contractual and technical requirements are met.

Supporting activities and system reliability

Supporting activities play a critical role in enabling reliable operations. ISO 9001 includes requirements related to competence, awareness, communication, infrastructure, work environment, and documentation. These elements ensure that operations are not dependent on individuals alone but are supported by a shared understanding, stable arrangements, and accessible information.

The control of documented information is part of this support structure. The purpose is to ensure that reliable information is available when and where it is needed to support consistent operation, learning, and continuity.

External providers and the extended value chain

ISO 9001 recognises that value creation often extends beyond the organisation's formal boundaries. Most organisations rely on suppliers, subcontractors, and service providers, whose performance influences the final outcome. Therefore, the standard requires organisations to control externally provided processes, products, and services in proportion to their impact on quality.

Some suppliers influence the final product or service directly, whereas others have a more indirect impact through supporting activities. Both can affect an organisation's ability to meet customer expectations and therefore require appropriate oversight. The greater the influence of an external provider on an organisation's ability to meet requirements, the more systematic the control. This may involve supplier qualification, clear communication of requirements, performance monitoring and ongoing collaboration.

From planning to disciplined operation

Taken together, the planning and operation requirements of ISO 9001 translate the organisational intent into consistent and disciplined practice. By identifying value-adding and supporting activities, defining their interactions, and managing them systematically, organisations create the conditions for consistent performance and reliable outcomes.

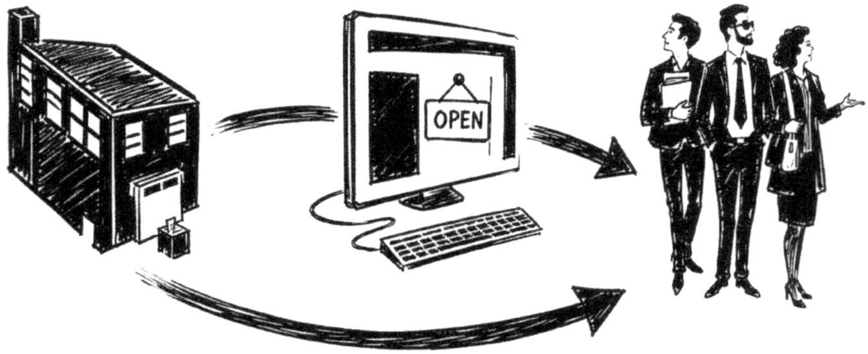

Figure 4.6 Direct and indirect influence of the suppliers on the customers.

This provides the foundation for performance evaluation, leadership oversight, and improvement.

4.6 Performance evaluation, leadership oversight, and continual improvement

Planning and disciplined operations are necessary conditions for consistent performance, but they are not sufficient on their own. Even well-designed processes may drift over time, encounter unforeseen conditions, or fail to deliver the intended results. ISO 9001 therefore places emphasis on performance evaluation, leadership oversight, and continual improvement as integral and recurring elements of the quality management system. Together, these elements ensure that the QMS remains effective, relevant, and aligned with the organisation's purpose and context over time.

Performance evaluation and learning

ISO 9001 requires organisations to systematically monitor, measure, analyse and evaluate their processes and outputs. The primary purpose of these activities is learning. By comparing planned arrangements with actual performance, organisations gain insight into how well their processes function and where adjustments are needed.

Customer satisfaction occupies a central role in performance evaluation, but it is only one of several dimensions that organisations must consider. Performance evaluation also includes process performance, conformity of products and services, and the effectiveness of controls.

The method of monitoring customer satisfaction varies depending on the nature of the organisation. A large consumer goods company may rely on surveys or market research based on representative samples, whereas a small farm selling specialised products to a limited number of customers may gather feedback directly through personal contact.-What matters is not the method itself, but that the organisation understands how

Figure 4.7 Farmer, getting direct feedback from his customer.

customers perceive the quality, reliability, and value of its products or services, and uses this information as input for improvement.

Internal audits provide a structured assessment of whether the QMS is implemented as intended and whether it remains effective. Audits should be understood as a learning mechanism, not as policing, and they are most valuable when used to identify weaknesses, inconsistencies, and improvement opportunities.

Data generated through monitoring, measurement, audits, and feedback must be analysed in a systematic manner in order to understand performance and identify improvement opportunities. ISO 9001 does not prescribe specific analytical methods but requires decisions to be based on evidence rather than intuition or isolated incidents.

Management review and leadership oversight

ISO 9001 assigns explicit responsibility to top management to oversee the quality management system. The management review is the formal mechanism through which this responsibility is exercised. Through regular and structured reviews, top management assesses whether the QMS remains suitable, adequate, and effective, and whether it continues to support the organisation's strategic direction.

Management reviews draw on information such as audit results, customer feedback, process performance, nonconformities, corrective actions, and changes in internal and external conditions. Active leadership involvement is essential; without it, performance evaluation risks becoming a routine exercise with limited impact.

In small organisations, management review is about structured reflection by management—based on factual information—on whether the system is working and what needs attention.

Nonconformities, corrective action, and improvement

When deviations from the requirements occur, ISO 9001 requires organisations to respond in a structured manner. Nonconformities must be identified, controlled, and corrected; however, the standard places equal emphasis on addressing root causes to prevent recurrence. Corrective actions are closely linked to organisational learning.

Beyond corrective actions, ISO 9001 promotes continual improvement. Improvements include incremental adjustments, more substantial changes to organisational arrangements, and proactive efforts to enhance performance and adaptability. In practice, many improvement initiatives are organised as projects with defined objectives, responsibilities, and timelines.

Continual improvement includes proactive efforts to reduce risk and strengthen organisational capability before problems arise. A simple example may help to illustrate this. Consider a small kindergarten in a rural town. The kindergarten may be functioning well, with no major quality problems, satisfied parents, and stable daily operations. However, management may become aware that a couple of experienced teachers have discussed the possibility of moving away from the area, perhaps because of temporary economic uncertainty or lack of professional development opportunities. If this occurs, the quality and continuity of care could be seriously affected. Rather than waiting for resignations to occur and then reacting, the kindergarten may choose to act proactively. By engaging in dialogue with its teachers, understanding their concerns, and exploring measures that could increase retention, such as flexible arrangements, development opportunities, or clearer future prospects, the organisation is acting before nonconformity occurs.

This type of proactive improvement demonstrates how continual improvement can be used not only to correct deficiencies, but also to anticipate risks, strengthen resilience, and support long-term quality in everyday operations.

Dynamic system

Performance evaluation, leadership oversight, and improvement ensure that the QMS remains a living system and evolves through feedback, reflection, and deliberate action. This dynamic character is essential in environments characterised by change, uncertainty, and increasing expectations regarding quality, responsibility, and transparency.

ISO 9001 does not guarantee perfection or stability. However, by embedding evaluation and improvement into regular management practice, the standard supports resilience and long-term organisational capability, enabling the organisation to remain stable in purpose while adapting to changing conditions.

4.7 Environmental management and ISO 14001

Environmental management has become an important topic for many organisations, driven by regulatory requirements, stakeholder expectations, and increasing awareness of environmental impact. ISO 14001 is an international management system standard that focuses on environmental management. Like ISO 9001, it is a generic standard that can be applied to organisations of different sizes and sectors, and it is designed to support systematic and disciplined management rather than prescribe technical solutions.

ISO 14001 is built on the same aligned management system framework as that of ISO 9001. It follows the same overall logic and structure and relies on familiar elements such as leadership commitment, planning, operational control, performance evaluation, management review, and continual improvement. For organisations that already operate a QMS in accordance with ISO 9001, much of the management system infrastructure required by ISO 14001 is already in place. The difference lies not in how the system is managed, but in what it is primarily concerned with.

ISO 14001 is distinguished from ISO 9001 by its explicit focus on environmental aspects and performance. The standard requires organisations to identify and assess how their activities, products, and services interact with the environment and to determine which of these interactions are significant. This analysis often extends beyond what is immediately visible in daily operations and may include indirect environmental impacts throughout the value chain.

A simple example illustrates this point. Consider an engineering consultancy whose primary output consists of drawings, specifications and advisory services. At first glance, the organisation's environmental impact may appear limited to office-related factors such as paper use, energy consumption, and employee travel. However, a closer analysis reveals that the consultancy's most significant environmental influence arises indirectly through the design choices it recommends to clients. Decisions regarding materials, energy solutions, and building layouts may have long-term environmental consequences that far exceed the consultancy's operational footprint. ISO 14001 encourages organisations to identify such less obvious impacts and to consider how their professional activities influence environmental outcomes beyond their immediate boundaries.

Figure 4.8 Implementing the ISO14001 standard calls for shedding light on all environmental aspects of the operation.

Based on the identification and evaluation of environmental aspects, organisations are expected to establish an environmental policy and define objectives that reflect both their environmental priorities and applicable legal and regulatory requirements. Operational controls and procedures are then adapted to ensure that activities are carried out in line with this policy. Monitoring and measurement focus on relevant environmental indicators, enabling organisations to assess performance, demonstrate compliance, and identify opportunities for improvement.

ISO 14001 builds on the same management system logic as ISO 9001, but applies it to environmental performance. Rather than introducing a new way of managing, it extends the existing framework to address environmental aspects in a structured and systematic manner. This allows organisations to integrate environmental considerations into their overall management system, using the same principles of planning, control, evaluation, and continual improvement. In this way, environmental management becomes part of a coherent and integrated approach to organisational performance.

4.8 Occupational health and safety and ISO 45001

Occupational health and safety are another area where organisations may choose to apply a dedicated management system standard. ISO 45001 is an international standard for occupational health and safety management systems. Its primary purpose is to help organisations prevent work-related injuries and ill health and to provide safe and healthy working conditions through systematic management.

ISO 45001 follows the same aligned management system framework as ISO 9001 and ISO 14001. Organisations that already operate a QMS will therefore recognise its overall structure and logic. As in the other standards, the difference lies in the specific focus of attention—in this case, on occupational health and safety risks.

Figure 4.9 Implementing the ISO45001 standard calls for shedding light on all occupational health and safety aspects of the operation.

The defining feature of ISO 45001 is its systematic focus on hazard identification and management of health and safety risks. While many organisations initially associate occupational health and safety with physical hazards, such as machinery, stairs, handrails, lighting, or ergonomics, the scope of the standard is broader. It explicitly requires organisations to consider all types of risks that may affect workers' health and well-being.

The engineering consultancy introduced in the previous section provides a useful illustration of this. At first glance, the physical working environment of such organisations may appear relatively safe. The office is well-designed, safety regulations are met, and there are few obvious physical hazards. However, a closer examination may reveal other less visible risks. Consultancy may be heavily involved in demanding projects with tight deadlines, high responsibility, and never-ending pressure from clients. Over time, this may lead to excessive workload, stress, fatigue, and risks related to mental health, burnout, and long-term absence.

ISO 45001 encourages organisations to identify and address such risks in a structured manner. This may involve examining workload distribution, project planning practices, availability of recovery time, clarity of roles, and the degree of support provided to employees. By engaging employees in dialogue and involving them in identifying risks and improvement measures, organisations can take preventive action before problems materialise.

Another characteristic of ISO 45001 is its strong emphasis on worker participation and consultation. Employees are a key source of insight into how work is actually performed and where pressures and hazards arise. By involving them in identifying risks and improvement opportunities, organisations gain a more accurate understanding of their operations and can take more effective preventive action. This emphasis reinforces the broader quality management principle of engagement of people and highlights the social dimension of systematic management.

ISO 45001 does not introduce a fundamentally new management philosophy. It applies the same principles of planning, control, evaluation, and improvement to a specific and critical area of organisational responsibility. Its close alignment with ISO 9001 enables organisations to address quality, environmental, and occupational health and safety concerns within an integrated management framework, while recognising that each area focuses on different types of risks—for example, product and service quality, environmental impact, and risks to human health and safety.

On a final note, for many organisations, environmental and occupational health and safety concerns provide a more tangible entry point to systematic management than quality in the abstract. Addressing these areas can, in some cases, make the value of structured processes, risk awareness, and continuous improvement more immediately visible to both managers and employees.

4.9 Excellence models

International standards are not the only reference that organisations can use when developing systematic quality management. In addition to standards, several quality management frameworks aim to support organisational learning, reflection, and long-term improvement. These frameworks differ fundamentally from certifiable standards in that they do not define formal requirements or lead to certification. Instead, they provide holistic models that organisations can use to assess their current situation, identify their strengths and weaknesses, and guide improvement efforts.

One of the most widely used quality management frameworks in Europe is the EFQM Excellence Model (European Foundation for Quality Management [EFQM], 2026)[12]. The EFQM model was developed to help organisations improve their overall performance and competitiveness by encouraging structured self-assessment and organisational learning. The model emphasises understanding how different elements of the organisation interact to create value and deliver results over time.

At the core of the EFQM model is a systems perspective built around three fundamental dimensions that organisations are encouraged to address systematically: direction, execution, and results. By explicitly linking purpose, culture, execution, and results, excellence frameworks make the deeper strategic and motivational drivers of systematic management visible.

The first dimension is direction, which focuses on purpose, strategy, leadership, and organisational culture. Organisations are encouraged to reflect on their fundamental reason for being, the value they aim to create for stakeholders, and how this is translated into a strategic direction. Leadership plays a central role in this dimension. Leaders are expected to provide clarity of purpose, set direction, and act as role models for values and behaviour. Organisational culture is treated as an integral part of direction, reflecting shared assumptions, norms, and ways of working that influence how strategies are interpreted and enacted.

The second dimension is execution, which focuses on how the organisation delivers results in practice. This includes how stakeholders are engaged, how partnerships and resources are managed, and how processes and activities are designed and coordinated to create sustainable value. Execution is broader than operational efficiency alone. It encompasses an organisation's ability to work coherently across boundaries, adapt to changing conditions, and balance short-term performance with longer-term development and transformation. Attention is paid to how consistently and effectively different parts of the organisation work together, and how organisational practices contribute to overall performance, resilience, and learning.

The third dimension of the EFQM model is results, which addresses what the organisation has achieved or plans to achieve. The results are viewed from multiple perspectives, including strategic and organisational performance, as well as the perceptions and experiences of key stakeholders. The model assumes that sustainable results are not accidental but the outcome of sound direction and effective execution of the project. By systematically examining results, organisations can identify patterns, understand cause–effect relationships, and use this insight to improve leadership practices, strategy, and execution. Therefore, the results serve as a basis for reflection and organisational learning rather than as an endpoint in themselves.

Currently, the EFQM model places a stronger emphasis on sustainable performance improvement. The concept of sustainable value refers to the organisation's ability to maintain long-term viability and high performance while simultaneously addressing environmental and social sustainability. In addition, greater attention is given to technology leadership, recognising that organisations increasingly depend on digital technologies and data-driven innovation to achieve strategic goals and deliver value to stakeholders.

In practice, organisations typically begin using the EFQM model through structured self-assessment. This process involves analysing the organisation against the different dimensions of the model, identifying strengths, weaknesses, and opportunities for improvement. Many organisations use the results internally as a basis for organisational learning and strategic development. Others take the process further by applying for external recognition programmes such as the European Quality Awards. In these cases, the organisation submits documentation describing its practices and results, and independent assessors provide detailed feedback following an evaluation and site visit. The intention of this process is not only recognition but also learning. Organisations are encouraged to reflect critically on their performance and use the feedback to strengthen their direction, execution, and results over time.

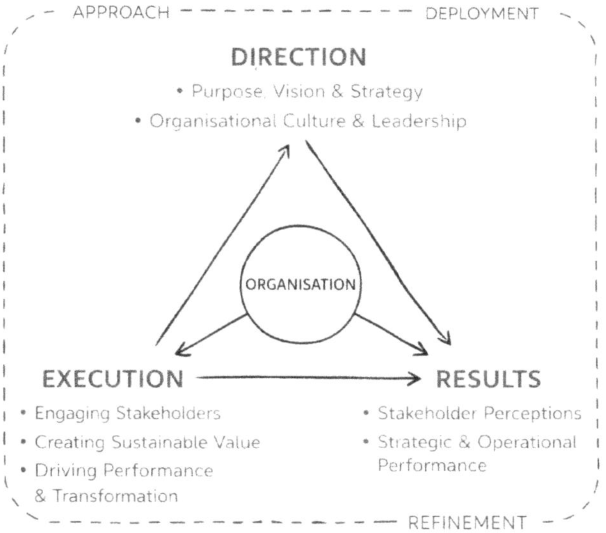

Figure 4.10 The EFQM model—direction, execution, and results (author's interpretation based on the official framework).

Figure 4.11 Winning international excellence awards.

A simple example can help illustrate why an organisation might choose EFQM as its primary reference instead of ISO 9001. Consider an international humanitarian aid organisation operating in regions affected by conflict, natural disasters, or political instability. Such an organisation must collaborate with a wide range of stakeholders, adapt rapidly to changing conditions, and balance accountability, ethical responsibility, and societal impact. In this context, a certification-oriented standard focused on the consistency of predefined processes may offer limited guidance. Instead, the organisation needs a framework that supports reflection on purpose, leadership, culture, partnerships, impact, and learning across highly variable and unpredictable situations. The EFQM model provides a holistic structure for such reflection without imposing a compliance logic that may be difficult to apply meaningfully in this type of operating environment.

This does not imply that EFQM and ISO 9001 are mutually exclusive. Many organisations use ISO 9001 to establish operational discipline and transparency while using EFQM as a broader reference for organisational maturity, performance, and long-term development.

Closely related to the EFQM model is the Common Assessment Framework or CAF (European Public Administration Network [EUPAN], 2026)[13], which applies a similar excellence-oriented logic to public sector organisations. CAF was developed by European public administrations specifically to support systematic performance improvement in government institutions. Unlike many other frameworks, CAF is free of charge and publicly available, which has contributed to its widespread use across municipalities, government agencies, and other public organisations. The framework retains the core idea of linking leadership, strategy, people, partnerships, resources, and processes to organisational results, but it adapts these concepts to the realities of public administration, where accountability, transparency, and societal impact often play a central role. In practice, CAF is primarily used as a self-assessment and organisational learning tool, helping public organisations identify their strengths, weaknesses, and priorities for improvement.

Outside the European context, the Malcolm Baldrige Framework provides a comparable, excellence-oriented approach to quality management. Although developed in a different institutional setting, it shares many similarities with the EFQM in its focus on leadership, strategy, customers, measurement, workforce, and results. Mentioning such frameworks illustrates that the underlying ideas of excellence, learning, and systematic improvement are not confined to a single region or tradition.

Quality management frameworks can be particularly useful where organisations wish to encourage reflection and development without the formality of certification or where the primary objective is organisational learning rather than external assurance. They can also complement the use of standards by broadening the perspective from compliance and control to performance, results, and organisational maturity. Frameworks such as the EFQM and CAF provide valuable reference points for organisations seeking to improve continuously and better understand the effectiveness of their quality management practices.

In some sectors, specialised control frameworks are also used. One example is the Hazard Analysis and Critical Control Points (HACCP) system in food production, which focuses on identifying and controlling biological, chemical, and physical hazards through the preventive monitoring of critical control points. Unlike general management system standards, HACCP concentrates on product safety rather than broader organisational management; however, its preventive logic has influenced many modern approaches to risk management and quality assurance. This methodology is incorporated into the ISO 22000 food safety management standard, which belongs to the same family of certifiable ISO management system standards and follows the same high-level structure discussed earlier in this chapter.

Case: Rebuilding a management system—which path forward?

The company in this case is a mechanical contracting firm that provides plant maintenance and installation services to heavy industry. It was established when a maintenance unit was separated from a larger industrial organisation and became an independent company. Several managers and employees had previously worked in a certified ISO 9001 system.

At first, this seemed like an advantage. Procedures existed, and the staff were familiar with audits and documentation. However, this transition revealed an important insight: a management system cannot simply be transferred from one organisation to another. The context has changed. The company was smaller, decision paths were shorter, resources were tighter, and client expectations were not identical. What worked in the parent company did not automatically fit the new organisation.

The company decided not simply to copy the previous system, but to rebuild its own. Customer expectations were an important driver. Instead of aiming immediately for formal certification, management chose to use ISO 9001 as a reference to understand and structure its own operations, and to guide discussions about responsibilities, processes, and risk.

The development was gradual. Managers worked closely with employees to describe actual work practices rather than designing ideal procedures. Some feared the emergence of a bureaucratic paper monster, but documentation soon proved useful in clarifying responsibilities and reducing misunderstandings.

The most visible improvements occurred in risk management and safety. Deviations were systematically documented, corrective actions were taken promptly, and formal risk assessments became routine. Over time, the company achieved a period with no reported accidents at work. This strengthened trust with clients and became a point of internal pride.

As the system matured, a strategic question emerged. The company recognised that it could not implement a ready-made solution from its former parent organisation. It now had to decide on the direction to take.

Several options were under consideration:

- Continue towards formal ISO 9001 certification.
- Shift focus towards ISO 45001, given the strong safety profile and "no accidents" ambition.
- Consider ISO 14001 in response to increasing environmental expectations.
- Or adopt a broader excellence framework, such as the EFQM, using it as a developmental reference rather than pursuing certification.

Each alternative implies different priorities, costs, and signals for stakeholders.

The company learned that standards provide structure, not answers. The next step would depend on how it interprets its risks, strategic ambitions, and the expectations of its clients and employees.

Reflection points—Chapter 4

1. Choosing whether to use a standard
   Think of a specific organisation you know well (for example, a regional construction firm, a private healthcare provider, or a university department) that is considering formalising its management practices. What concrete problems or stakeholder pressures would justify adopting an international standard, and what risks might arise if the initiative is driven mainly by image or competitor behaviour? Under what conditions would a non-certifiable framework be more suitable than ISO certification?
2. Diagnosing the PDCA balance in practice
   Choose a real organisation or project you are familiar with, such as a municipal service unit or an engineering consultancy, and analyse it using PDCA logic. Where is planning strong but evaluation weak, or where are corrective actions taken without systematic learning? What operational or strategic risks emerge if one phase of the cycle is underdeveloped?
3. One system or fragmented responsibilities
   Consider a mid-sized manufacturing company that already applies ISO 9001 and is now facing increasing safety and environmental expectations. Should it extend its system in

a coordinated manner or allow different departments to manage quality, environment, and safety separately? What risks arise if responsibilities, audits, and reviews are fragmented across parallel structures? How might early integration choices influence long-term organisational coherence and maturity?

4. Proactive improvement before failure
   Consider a small kindergarten, consultancy, or logistics company that currently experiences few visible problems. How can leadership justify investing in preventive improvements before nonconformities occur? What risks arise if improvements are triggered only by accidents, complaints, or regulatory pressure?
5. Seeing environmental impact beyond the obvious
   Consider a service-based organisation, such as an architectural firm or IT consultancy, that assumes its environmental footprint is limited to office energy use. How could a value chain perspective reveal the indirect environmental consequences of design choices or recommendations? What tensions might emerge between cost efficiency, client demands, and long-term environmental responsibilities? How could ISO 14001 reshape decision-making priorities, rather than simply introducing new procedures?
6. Certification or excellence orientation
   Reflect on an international humanitarian aid organisation operating in unstable environments. Would ISO 9001 certification primarily strengthen accountability and external trust, or would it constrain adaptability? How could an excellence framework focus on direction, execution, and results support a different governance logic?

Notes

1 International Organization for Standardization. (2026). ISO 9001:2026 Quality management systems—Requirements (ISO Standard No. 9001:2026).
2 International Organization for Standardization. (2015). ISO 9000:2015 Quality management systems—Fundamentals and vocabulary (ISO Standard No. 9000:2015).
3 International Organization for Standardization. (2015). ISO 9001:2015 Quality management systems—Requirements (ISO Standard No. 9001:2015).
4 International Organization for Standardization. (2018). ISO 9004:2018 Quality management—Quality of an organization—Guidance to achieve sustained success (ISO Standard No. 9004:2018).
5 International Organization for Standardization. (2025). ISO 19011:2025 Guidelines for auditing management systems. (ISO Standard No. 19011:2025).
6 International Organization for Standardization. (2015). ISO 9001:2015 Quality management systems—Requirements (ISO Standard No. 9001:2015)..
7 International Organization for Standardization. (2026). ISO 14001:2026 Environmental management systems—Requirements with guidance for use (ISO Standard No. 14001:2026).
8 International Organization for Standardization. (2015). ISO 45001:2018 Occupational health and safety management systems—Requirements with guidance for use (ISO Standard No. 45001:2018).
9 International Organization for Standardization. (2022). ISO 27001:2022 Information security, cybersecurity and privacy protection—Information security management systems—Requirements (ISO Standard No. 27001:2022).
10 International Organization for Standardization. (2018). ISO 22000:2018 Food safety management systems—Requirements for any organization in the food chain (ISO Standard No. 22000:2018).

11 International Organization for Standardization. (2018). ISO 50001:2018 Energy management systems—Requirements with guidance for use (ISO Standard No. 50001:2018).
12 European Foundation for Quality Management. (2026). EFQM. Retrieved May 20, 2026, from https://www.efqm.org/.
13 European Public Administration Network. (2026). CAF. Retrieved May 20, 2026, from https://www.eupan.eu/caf/.

5 Implementing quality management

A project approach

5.1 Implementing a QMS as a project

Implementing a quality management system is a deliberate and structured effort to shape an organisation's operations. It involves defining responsibilities, clarifying expectations, agreeing on ways of working, and establishing mechanisms for follow-up and improvements. There is a clear starting point, defined objectives, limited resources, and a need for coordination across organisational units. In other words, the implementation of a QMS can be understood as a project.

A project perspective helps to make the specific challenges of such efforts visible. One of the most important challenges is that the implementation involves organisational change. The introduction of a QMS affects everyday work practices and requires time and attention from both managers and employees. Therefore, it competes with ongoing operations and other priorities. Treating implementation as a project provides a practical way of organising this effort, clarifying roles and responsibilities, and maintaining focus on the overall purpose.

Experience from organisations that have implemented QMSs and undergone ISO 9001 certification supports this view. In practice, most organisations claim that they

Figure 5.1 The QMS implementation project.

DOI: 10.4324/9781003671749-5

see the implementation as a project and organise it accordingly, with defined phases, milestones, and some form of planning and follow-up. However, claiming that project management is being applied does not necessarily mean that it is being done in a meaningful way. Project management is more than just phrases and simplistic tools. It is about understanding the specific challenges of each project and planning and executing the work accordingly. The way the implementation is approached has a decisive influence on the outcome.

QMS implementation should be viewed as a particular type of project, a change management project. Its primary outcome is not a physical product, but changes in how people work and how the organisation coordinates its activities. Procedures, records, and roles are introduced, but their value depends on whether they are understood and used in practice. Leadership, communication, and employee participation play central roles in implementation.

This chapter builds on this perspective. Following the logic of a project lifecycle, it discusses QMS implementation as a sequence of connected phases, from the initial strategic decision to the transition into normal operation. The intention is to provide a way of thinking about implementation that reflects practical experience and supports organisations that decide to invest seriously in quality management.

5.2 The QMS implementation lifecycle: Phases, tasks, and milestones

When implementing a QMS, it is essential to maintain an overview of the entire effort while simultaneously recognising that different types of work are required at different points in time. A useful way to think about this is through an analogy with air travel. When flying at a high altitude, the landscape below appears as a coherent whole. Major structures are visible, whereas details are deliberately omitted. As the aircraft descends, more details gradually come into view, and attention can be shifted to specific features on the ground. The implementation of a QMS follows a similar logic in terms of content and over time.

This logic can be described using the concept of a project lifecycle. This is a simple way of structuring the implementation effort into phases, each characterised by a particular focus and type of activity. In the context of QMS implementation, it is useful to distinguish between four main phases: concept, planning, execution, and delivery with transition to operation. These phases do not represent rigid steps or a strictly sequential process, but they reflect how the nature of the work changes as implementation progresses. A useful way to visualise this lifecycle is through a simple Gantt chart:

Figure 5.2 Steps in the implementation of QMS, from the general to the specific.

Each phase consists of a set of tasks. In the early phases, tasks are primarily concerned with direction, commitment and understanding. Subsequently, the emphasis shifts towards developing, applying, and testing the system in practice. Finally, tasks focus on reviewing, learning, and preparing the organisation to operate the system as an integrated part of normal management. Although these phases are presented separately for clarity, they often overlap and interact in practice. The lifecycle model should be understood as a logical structure, rather than a rigid sequence. For instance, planning may continue while execution has already begun, and learning from early application may influence both the objectives and detailed design.

Between phases, there are typically milestones when something important has been achieved or clarified, a decision has been made, or something has been delivered. In a QMS implementation project, such milestones may include the formal decision to proceed, approval of the quality policy and objectives, readiness to begin internal audits, and completion of certification if that is an objective. Milestones can be natural moments for reflection, decision-making, and confirmation of commitment by the top management.

The flight analogy helps illustrate how this life cycle should be approached in practice. In the concept phase, the organisation is "flying high," focusing on purpose, direction, leadership, and overall structure. During the planning phase, the organisation descends somewhat, developing a clearer picture of its current situation and preparing for implementation. In the execution phase, the QMS is gradually built and put into use step by step. Procedures are developed, tested in practice, and improved based on experience, while employees become familiar with new ways of working. In the delivery phase, the focus shifts towards verifying that the system works as intended and ensuring that the new procedures are integrated into everyday operations so that the organisation moves from project mode to normal management.

In practice, the boundaries between these phases are often blurred. Execution and delivery often overlap as parts of the system are developed, applied, and stabilised while other parts are still being built. Learning from early application may also influence both planning and further development, making implementation an iterative process rather than a linear one.

A common mistake in QMS implementation is to focus too early on the detailed documentation. The result may be extensive technical documents that are not needed or used. The purpose of developing procedures, work descriptions, and technical guidelines

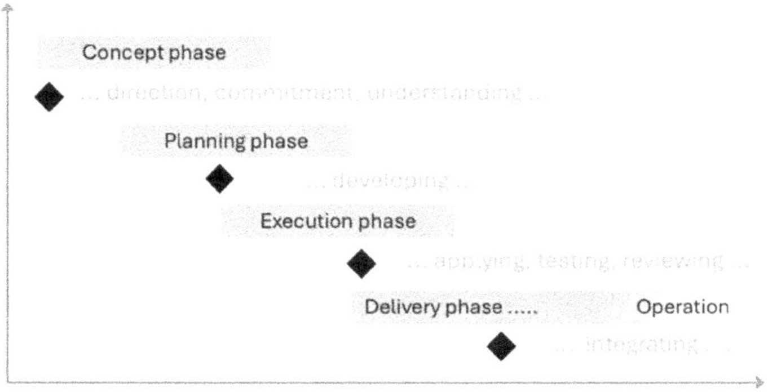

Figure 5.3 QMS implementation shown as a Gantt chart.

is to help the organisation meet defined expectations in a consistent and reliable manner. Documentation should be developed to add clarity and support to the operation. If expectations can be met without detailed technical documentation, then producing such documentation may have little value.

The lifecycle perspective helps distinguish between implementation as a project and operations as an ongoing system. The implementation project is temporary and has a defined endpoint, perhaps when the QMS is fully operational or even when a formal certification has been issued. At that point, the responsibility shifts from the project organisation to line management and the organisation's permanent governance structures.

Making this transition explicit reduces the risk of the QMS being perceived as finished and final once the project ends. Instead, it reinforces the understanding that the QMS is a living system that requires continuous attention, follow-up, and improvement as part of the normal management.

The subchapters that follow are structured according to this lifecycle, moving from high-level conceptual considerations to more detailed aspects of planning, execution, and delivery.

5.3 Concept phase: The strategic decision to invest in quality management

The implementation of a QMS begins with a strategic decision. This decision is typically made at the top management or board level and reflects a deliberate choice to invest in quality management to shape the future of the organisation. It is a high-level decision about how the organisation intends to operate, coordinate its activities, and meet expectations systematically. It must be based on realistic expectations regarding the outcome, a better coordinated and more focused operation that can lead to improved products or services at a lower cost.

This is particularly important in organisations where there is little tradition of formal work procedures. In such organisations, the implementation of a QMS can lead to considerable changes for employees. These changes may also extend beyond the organisation, influencing suppliers and, ultimately, the experience of clients.

This decision marks the entry point into the concept phase of the implementation lifecycle. The most important questions are why such a system should be built and what role it is expected to play in the future. The decision is often motivated by several factors simultaneously, such as customer requirements, competitive pressure, regulatory expectations, or a recognised need for better internal coordination. Regardless of the combination of drivers, top management must have a clear and shared understanding of the purpose of implementation.

Consider a public utility company that operates in several locations. The CEO repeatedly emphasises employee safety as a strategic priority in management meetings and the importance of following the safety manual. Three of the four department heads take this seriously and reflect it in their daily decisions; one listens but does not treat safety as a real priority. The difference is quickly felt further down the organisation. Middle managers who report to aligned department heads receive a clear and consistent message. Those who report to the fourth received mixed signals. Over time, this leads to uneven practices and confusion among the employees. The lesson is simple. A strategic priority only becomes real when the management team is aligned. Without such alignment, systematic initiatives, such as the implementation of a QMS, are weakened from the start.

Figure 5.4 The managers lead the way in the implementation of a QMS.

Making this strategic decision also means accepting that the organisation is committing itself to change. Formalising work practices, clarifying responsibilities, and introducing systematic follow-up will affect everyday work. These decisions cannot be delegated or treated as minor steps. Without clear ownership by top management, the implementation project loses direction and credibility at an early stage.

Another important aspect concerns the use of employee time. The time spent by employees on activities such as mapping processes, standardising ways of working, and participating in training represents a real and significant commitment. During this period, employees do not create direct operational value. The expectation is, of course, that this investment will pay off, for example, through reduced waste, better utilisation of resources, improved service quality, and greater satisfaction among clients and staff. However, these benefits may take time to materialise and must be understood as part of strategic commitment.

An important outcome of the concept phase is therefore explicit commitment, a willingness by top management to prioritise implementation, allocate the necessary resources, and remain actively involved. Additionally, communicating the decision clearly within the organisation so that employees understand both the intention and the expected implications.

Finally, the strategic decision to invest in quality management establishes the first major milestone of the project, namely, the decision to proceed. This is a reference point for subsequent planning and execution and signals that the organisation is ready to move from intention to action. The implementation can progress towards more concrete considerations of leadership, governance, and organisational readiness. In many organisations, the decision to implement a QMS is also linked to maintaining trust and legitimacy among customers, regulators, and society at large. Systematic quality management signals reliability and responsible governance beyond immediate operational concerns.

5.4 Concept phase: Leadership and governance as preconditions

Once a strategic decision is made, attention must be paid to leadership and governance. Without clear leadership and an appropriate governance framework, even a well-intentioned QMS initiative is likely to lose its momentum. However, when leadership is aligned, quality management becomes an expression of what the organisation stands for.

Leadership refers to the active and visible involvement of top management. Top management is responsible for quality management, and this overall responsibility must be demonstrated in practice and cannot be delegated. This includes showing interest, asking questions, participating in key discussions, and acting as role models by working in accordance with the introduced principles. Employees quickly recognise whether quality management is taken seriously.

Governance concerns how responsibilities are clarified, decisions related to implementation are made, and follow-up is ensured. The key question is whether the existing organisational arrangements can support the implementation effort or whether some changes are needed to make quality management visible and credible.

Many organisations choose to establish a coordinating arrangement, sometimes referred to as a quality committee or steering group, to support top management during implementation. The purpose of this is to ensure coordination, prioritisation, and follow-up across organisational units. In smaller organisations, top management may directly handle these responsibilities without formal structures.

It is also important to clarify the role of those who coordinate implementation on a daily basis. This is a crucial step in every project, where an appropriate project organisation must be established. In project management terminology, typical roles include a project manager who facilitates and coordinates the implementation work and a project sponsor or steering group to whom the project manager reports. In many organisations, the person responsible for coordinating the implementation is formally appointed as a quality manager. During the implementation phase, however, this role largely corresponds to that of a project manager, focusing on coordination, communication, and

Figure 5.5 Leadership in action, "walking the talk."

follow-up. These roles support cooperation and help translate strategic intent into practical action. However, responsibility for quality remains with management and employees throughout the organisation.

Middle managers play a particularly important role. They connect leadership intentions with operational reality and strongly influence how employees perceive the implementation effort. Their engagement, or lack thereof, often determines whether the QMS is seen as a meaningful management initiative or an external imposition.

A more detailed discussion of the organisational structure, roles, communication, and information flows is provided in Chapter 7.

5.5 Understanding the initial situation

A fundamental requirement in any project, particularly in organisational change projects, is to develop a solid understanding of the initial situation. This is a precondition for realistic planning, credible scheduling, and meaningful resource allocation. This phase involves clarifying the nature of the project and the challenges ahead. The implementation of a QMS is not the same in all organisations. The effort required depends on factors such as existing ways of working, the degree of formalisation already in place, the size and complexity of the organisation, and the expectations of employees, customers, regulators, and other interested parties. Gaining an overview of these conditions helps to understand what kind of project lies ahead.

A common starting point is simple gap analysis. This involves comparing current practices with a chosen reference, such as a management standard or a quality framework. The purpose is to identify where current practices are aligned with this reference and where significant gaps exist. This becomes a valuable input for defining priorities, sequencing activities, and deciding where efforts should be concentrated.

Understanding the starting point also involves clarifying the project scope. A key question is whether the QMS is intended to cover the entire organisation or only parts of it. In some organisations, it may be reasonable to start with a single unit or function and expand later, while in others, the nature of governance and support processes makes this impractical. The organisation should also consider how its processes interact with suppliers, customers, and other external actors, as implementation decisions may influence value creation beyond organisational boundaries.

To understand the initial situation, it is also useful to apply basic tools such as stakeholder and risk analyses. Stakeholder analysis helps identify individuals and groups who are affected by the implementation or who can influence its outcome, such as employees, managers, customers, suppliers, or external bodies such as regulators, certification bodies, or relevant public authorities. Risk analysis helps surface uncertainties related to resistance to change, lack of resources, competing priorities or unrealistic expectations. These analyses do not need to be complex, but they help management anticipate challenges rather than react to them later. The implementation can be designed in a more informed manner, increasing the likelihood of success.

Issues related to organisational culture and quality culture are also highly relevant when assessing the starting point, as they influence how change initiatives are perceived and adopted. At this stage, however, the purpose is primarily to acknowledge their importance and consider them in planning. A more detailed discussion of organisational culture, quality culture, and their roles in quality management is provided in Chapter 7.

The main outcome of this phase is a shared and realistic understanding of the organisation's current status and what the implementation project is likely to involve. This understanding forms the basis for defining the objectives, scope, and approach of the planning work that follows. By understanding the initial situation, the organisation improves its ability to plan implementation in a way that is both feasible and aligned with its context.

5.6 Planning phase: Defining project goals and quality objectives

At this stage, the organisation must clarify where it wants to go and what the goals and objectives of the implementation project are. This task is inseparable from strategic thinking and from shaping the organisation's quality policy and quality objectives. This is the point in the implementation lifecycle where strategy and project management meet most clearly.

Every project begins with an idea of the future it intends to create. The same applies to the implementation of the QMS. Before deciding how the system should be developed and implemented, an organisation must first clarify what it wants to achieve through quality management.

This logic is reflected in a quality pyramid. At the top of the pyramid is the quality policy, which expresses the organisation's overall intentions regarding quality. The quality policy provides direction and answers fundamental questions about how quality is understood and prioritised within the organisation. Shaping the quality policy is the responsibility of top management and forms a central part of strategic planning.

When shaping the quality policy, some basic questions should be addressed. In particular, how the needs and expectations of customers and other interested parties are assessed and understood. A clear commitment to continuous improvement should be

Figure 5.6 The quality pyramid is built from the top to the foundation.

emphasised and reflected in the organisation's culture, not only through formal statements. The quality policy should also reflect an understanding of the basic conditions for the organisation's existence: its clients, objectives, and core operations. When defining the policy, organisations may also consider broader responsibilities, such as safety, environmental performance, and long-term societal expectations. In this way, sustainability becomes part of how quality is understood and managed, rather than a separate concern.

Although these issues are part of strategic planning in a broad sense, they take on a more specific meaning in the context of quality management. Therefore, when defining a quality policy, it is also necessary to assess the organisation's status regarding quality. This may include considering quality-related costs, the organisation's position in the market in terms of quality, the prevailing quality culture, and how far the organisation has progressed in formalising its work procedures. It is also relevant to consider whether documented procedures and other quality-related documents already exist and how they are used in practice.

Quality objectives translate strategic intentions into more concrete priorities. Quality objectives address questions such as which quality-related topics are to be emphasised in the coming period and what measurable outcomes the organisation is aiming for. Some objectives may be directly related to operational performance, such as reducing defects, improving delivery reliability, or increasing customer satisfaction. Others may relate more closely to the implementation project itself; for example, achieving certification within a defined timeframe.

It is useful to distinguish between the project goals and quality objectives. The project goals concern what the implementation project is expected to deliver within a limited timeframe, such as putting a QMS in place and making it operational. Quality objectives, on the other hand, extend beyond the implementation project and relate to how the organisation intends to operate and improve over time. For example, consider a travel company that specialises in organising trips to international sports events. Let us assume that this company plans to implement a QMS and obtain international certification. Each trip can be seen as a project involving communication with clients, coordination of flights, accommodation, tickets, and local logistics. As part of introducing a QMS, the company can define practical quality objectives for the first year. One objective could be to begin systematically collecting data on estimated and actual project budgets to improve planning and financial control. Another objective could be to establish a more formal method for collecting detailed feedback from customers after each trip, as well as regular "lessons learned" meetings, where the team reviews completed projects and identifies improvements for future events.

Clarifying the relationship between quality policy, quality objectives, and project goals helps ensure alignment between long-term strategies and short-term implementation. It also helps avoid a narrow focus on certification or documentation as an end. Instead, the implementation project becomes a means of moving the organisation towards a defined future state.

Figure 5.7 provides indicates how quality policy and quality objectives are related and how project objectives relate to the project life cycle of implementing a QMS. With the quality policy clarified and project goals and quality objectives defined, the implementation project has a clear direction. This creates a solid foundation for the next step in the planning phase of the project.

Figure 5.7 Quality policy, quality objectives and project goals.

5.7 Planning phase: Project goals, scheduling, resources, and cost

At this stage, the task is to translate intentions into a realistic plan for how the implementation will be carried out. This means defining project goals and establishing a credible plan for time, resources, and cost. Planning should also consider digital platforms for documentation, communication, and data integration, as these platforms increasingly shape the application of quality systems in practice.

Project goals are now concerned with the success of implementation efforts. They may include reaching defined milestones, such as mapping the key processes of the organisation in a collaborative effort by the staff, completing one round of internal audits for the whole organisation, or going through an external audit within a given timeframe. Project goals may also relate to keeping the use of resources within agreed limits, covering a clearly defined scope, and ensuring that the implementation is manageable for the employees and managers involved. In addition, they may include stakeholder-related aspects, such as maintaining stable operations during implementation or seeking feedback from external stakeholders, whose perceptions may provide useful input to the process.

A key insight from empirical research on QMS implementation is that realistic planning of time and resource use is decisive for success. Studies of organisations implementing ISO 9001-based quality management systems show clear differences between organisations that treat implementation as a project with explicit planning and follow-up and those that rely mainly on informal coordination. Organisations that made a realistic plan for time and resource use reached their objectives faster and more predictably than those that did not (Ingason, 2015).[1]

The critical issue here is not primarily the cost of consultants or certification, but the time that employees must invest in the implementation. Implementing a QMS requires staff to engage in activities beyond their normal operational tasks. This includes developing a shared understanding of what quality means in the organisation, mapping processes, discussing problems, agreeing on standardised ways of working, and participating in training, audits, and improvement activities. All of this requires time, and during this time employees may not be producing direct operational output.

In fact, this investment in shared understanding and coordinated ways of working is one of the most important outcomes of a successful QMS implementation. If this is understood and acknowledged, it becomes possible to make a realistic estimate of

Figure 5.8 Realistic implementation means active participation of employees.

the resources required for implementation. However, if employees' time commitment is underestimated, implementation plans will be unrealistic. This is likely to lead to frustration and a loss of credibility.

From a project perspective, this reinforces the importance of linking time planning, resource allocation and cost estimation. Planning the implementation project is about creating a shared and realistic understanding of what the organisation is committing to. When time has been set aside for learning, discussion, and development, implementation is more likely to be perceived as a serious and credible initiative.

With project goals clarified and realistic plans for time, resources, and cost in place, the next step is to focus on developing and applying the QMS in practice.

5.8 Execution phase: Developing and implementing the QMS

This is the phase in which the QMS is developed, applied, tested, and gradually improved through its use. It involves coordinating activities and supporting people in translating intentions, policies, and objectives into practices that actually work in everyday operations.

A central task in the execution phase is to shape the structure of the QMS. This structure must make sense and be useful for users. Simultaneously, it must reflect any chosen reference framework, management standard, or quality model. Therefore, the result is a system that is both context-specific and aligned with external expectations. Finding this balance is part of the execution work and cannot be achieved purely through design on paper.

The development and use of the QMS can begin with a limited number of agreed-upon procedures or guidelines. They should include key processes, where coordination and consistency are important. Once a procedure is defined, it can be immediately put into use, tested in practice, and adjusted based on experience. The aim is to provide sufficient structure to support reliable performance and learning. This stepwise approach helps maintain an overview while gradually addressing relevant details and reduces the risk of producing documentation that adds little value.

Organisations should resist the temptation to begin implementation by producing large numbers of detailed technical descriptions. Experience shows that this approach often leads to excessive documentation and loss of overview. It is usually more effective to begin at a higher level by clarifying the organisation's purpose, identifying its key processes, and developing a limited number of general procedures. More detailed descriptions can then be introduced gradually where they are genuinely needed. Starting from the overall structure and moving step-by-step towards greater detail helps maintain strategic focus and prevents the QMS from becoming a documentation exercise.

Development and implementation are closely intertwined in this phase. Initial versions of procedures and guidelines are developed in a collaborative effort, issued, tried out in practice, discussed, and adjusted. The QMS is improved as it is implemented. When documenting procedures, it is important to start from the actual situation within an organisation. A common mistake is to begin by describing how work should ideally be carried out rather than how it is currently performed. This approach is risky because it attempts to introduce several changes simultaneously. A more effective approach is to begin by achieving a shared understanding of the current situation. This may be challenging, since there may initially be no fully standardised way of working. Reaching an initial common understanding of a procedure is therefore an important milestone, after which the team can begin discussing how the procedure can be improved over time.

Employees play a central role in this process. They are the ones who apply the QMS in practice and quickly discover what works and what does not. Through their experience, any weaknesses, ambiguities, or unnecessary complexities become visible. Their feedback is essential for improving the system. Participation in process discussions, improvement activities, and trials of new procedures can turn the QMS into a practical tool. Training and communication are closely connected to this participation. Communication must be

Figure 5.9 A mountain of unnecessary documents.

continuous and two-way, allowing questions, concerns, and criticisms to surface and be addressed. When employees understand why certain practices are introduced and see that their input leads to adjustments, their commitment to the system increases.

As implementation progresses, organisations often introduce internal checks or reviews to determine whether the QMS is functioning as intended. In ISO-based systems, internal audits are a formal requirement, but in other contexts, similar activities may take less formal forms. The underlying idea is the same: to step back, examine how things are done, identify weaknesses, and look for opportunities for improvement. Such self-reflection is a natural and necessary part of execution and learning.

The execution phase is both demanding and resource intensive. It coincides with ongoing operations and often with other changes in the organisation. This is precisely why a project perspective is valuable: it helps maintain focus, structure, and momentum during a period when fatigue and loss of attention are common risks.

5.9 Delivery and transition to operation

The final phase of the implementation project concerns the delivery and transition. This is the phase in which the implementation project is concluded, and responsibility is fully embedded in the permanent management structures of the organisation. The real objective is long-term and is expressed through the quality policy and the fundamental principles of quality management discussed earlier in this book. It is about coordinated ways of working, a focus on meeting expectations, learning from experience, and continuous improvement.

A key question is when the implementation can be reasonably considered complete. The implementation of quality management is a never-ending journey. However, the implementation project must have a defined endpoint. In some cases, it is clearly defined from the outset, for example, when achieving certification is set as a formal project milestone. In other cases, the endpoint may be defined in terms of internal criteria, such as having agreed and applied a set of core procedures, completed internal reviews, and conducted management reviews. What matters is that the organisation has a shared understanding of what marks the conclusion of the project and the transition to normal operations.

Consider a civil construction contractor that decides to build a formal QMS with the clear ambition of reducing accidents. The initiative is driven by internal values and increasing demands from major clients. ISO 45001 was chosen as a reference, and the implementation was organised as a project with a dedicated project manager and broad employee involvement. Within two years, the company achieved a 75 per cent reduction in reported accidents. From a safety perspective, this is a significant achievement. However, an important question remains: Is the implementation project finished? Certification was not the objective; safer work practices were the objective. Has safety become an integrated part of everyday work, or does it still depend on the project structure?

The vignette highlights a central challenge in this phase. While quality and safety management have no natural endpoint, the implementation project must still be concluded, with responsibility clearly transferred to normal management.

For organisations that choose to seek certification according to a management standard, such as ISO 9001, this phase includes preparation for and completion of an external audit. Certification can serve as a clear and visible milestone both internally and externally. It provides independent confirmation that the QMS meets the requirements of the chosen standard and has been implemented systematically. Simultaneously, certification should be understood as a milestone, not an endpoint. Its value depends entirely on how the system is used in practice after the auditors leave.

Internal audits and management reviews can play important roles in preparing for this transition. They provide structured opportunities to step back and assess whether agreed-upon ways of working are followed, whether the system supports the organisation's objectives, and where improvements are needed. Although internal audits are a formal requirement in ISO-based systems, the underlying logic is more general and reflects sound managerial practice. Organisations benefit from regularly examining their practices in a structured and critical manner.

Closing the implementation project involves acknowledging what has been achieved, clarifying remaining open issues, and confirming how the responsibility for maintaining and developing the QMS is integrated into everyday management. Without such closure, there is a risk that the implementation remains an ongoing "project" without clear ownership or that quality management is seen as something separate from normal managerial responsibility. Closing a project does not mean closing the door on improvement. On the contrary, the transition marks the point at which quality management has become embedded in how the organisation is managed. When successfully integrated, the QMS helps the organisation maintain direction and coherence, even during periods of change and uncertainty.

Continuous improvement, learning, and adaptation to changing expectations now occur within normal management processes rather than through dedicated projects. This ongoing development and the systematic pursuit of better performance over time are discussed later in the book.

The delivery phase brings the implementation project to a close, while simultaneously establishing the conditions for quality management as an integral and enduring part of organisational life.

Figure 5.10 Celebrating the closure of the implementation project and discussing what happens next!

Case: Implementing a QMS as a project in a logistics, courier, and package delivery company

The organisation is a national logistics, courier, and package delivery company in a European country, employing approximately 1,000 people across regional and local service units. The core challenge was to provide consistent services across geographically dispersed operations while responding to increasing expectations regarding efficiency, transparency, and service quality.

In response to the need to standardise work procedures and improve coordination, top management made a strategic decision to implement a quality management system. The intention was not merely to meet external expectations, but to establish a more systematic and transparent way of managing operations. From the outset, the implementation was framed explicitly as a project with a defined timeframe, clear milestones, and designated roles.

A quality manager was appointed to coordinate the implementation, which was supported by a project manager and a quality committee. External consultancy support was used in the early stages to structure the work and generate momentum. An implementation plan was developed with a target of 18 months—a demanding but realistic timeframe given the size and complexity of the organisation.

The time commitment of managers and employees was explicitly estimated. Middle managers were made responsible for developing and applying procedures in their own areas, while the implementation team provided structured support during the defined periods.

Implementation began with the organisation's core operational units, such as package sorting, delivery coordination, customer service, and handling of delivery deviations, followed by supporting functions. Rather than documenting everything in advance, the organisation focused on developing a limited number of key procedures, putting them into use, and improving them based on experience. The procedures were kept simple and described actual practice, for example, how delivery deviations were registered, how customer complaints were handled, and how communication between regional units was coordinated.

Internal audits were introduced after nine months and used as learning tools to identify inconsistencies, unclear responsibilities, and opportunities for improvement in the process. A pre-audit by the certification body provided additional input before the formal certification audit was conducted.

Certification was achieved within the planned timeframe, marking the formal conclusion of the implementation project. It was not treated as the primary objective, but as confirmation that the system functioned in practice.

After certification, the responsibility for running the QMS was integrated into the normal management structure. Procedures were made accessible through internal systems, and communication regarding performance and improvement was strengthened through regular meetings and visual displays.

The organisation observed tangible benefits. Employees gained clearer guidance, service consistency improved across regions, internal communication became more structured, and failure costs were reduced. Communication with suppliers and subcontractors has also become more systematic.

Based on the discussion in this chapter, which aspects of this organisation's approach stand out as especially important for the success of the implementation project?

Reflection points—Chapter 5

1. Alignment at the top
   In a public healthcare organisation, the executive team agrees to implement a QMS, but individual directors prioritise different objectives such as cost control, patient safety, and digitalisation. How might such divergence affect implementation, and how could this be dealt with? Discuss the trade-off between a strong central direction and managerial autonomy.
2. The scope of implementation
   A construction contractor with multiple regional units must decide whether to implement a QMS across the entire organisation or begin with one pilot unit. What are the risks and benefits of each approach in terms of learning, credibility, speed, and employee fatigue? How might governance and shared support functions influence the feasibility of partial implementation? Reflect on how this decision would differ in a highly centralised versus a decentralised organisation.
3. Employee time as investment
   In a knowledge-intensive consulting firm, employees already experience high workload pressures. The management plans a 12-month QMS implementation without reducing operational targets. What are the potential consequences? How should employee time be treated: as an overhead cost, strategic investment, or temporary inefficiency? Discuss the trade-off between short-term productivity and long-term capability building. At what point does underestimating the internal time commitment jeopardise the entire project?
4. Documentation depth and practical use
   A logistics company begins documenting procedures in detail before clarifying its quality policy and objectives. How would you assess the risk of documentation becoming detached from operational needs? Under what circumstances is detailed technical documentation necessary, and when does it add complexity without adding value? How can management ensure that procedures remain useful rather than ceremonial?
5. Certification milestone or strategic outcome
   A civil engineering firm achieved ISO certification within the planned timeframe, but the accident rates and customer complaints remained unchanged. Has the implementation project been successful? Distinguish between project success and quality performance. What indicators would you use to evaluate whether certification has led to more long-term value?
6. Project closure
   In a municipal service organisation, the QMS implementation project formally ends after certification, and the project team is dissolved. What conditions must be in place to ensure that quality management continues as part of normal management? How can responsibility be transferred without losing momentum? Discuss the tension between defining a clear project endpoint and recognising that quality management is an ongoing endeavour.

Note

1 Ingason, H.T. (2015). Best project management practices in the implementation of an ISO 9001 quality management system. *Procedia-Social and Behavioral Sciences*, 194, 192–200.

6 Understanding the activities of the organisation as a set of processes

6.1 Why a process approach?

Organisations are often described through their formal structure, departments, functions, reporting lines, and job titles. While such descriptions are useful for certain purposes, they say little about how work is carried out. Value is rarely created within organisational boxes alone, and customers are seldom served by a single department. Instead, value is typically created through work that flows across functions, roles, and organisational boundaries, often involving coordination between multiple actors.

A process approach offers a different and often more realistic way of understanding an organisation. It asks what happens, in what order, and who makes it happen. The focus is on sequences of activities that transform inputs into outputs, and on how these activities are linked together to create value for customers and other stakeholders.

This way of thinking plays right into the core of quality management. If a process has been mapped and it represents a shared understanding of that operation, the process can be controlled, and it can be improved. Many quality problems do not arise because individuals fail to perform their tasks, but because interfaces between activities are unclear, responsibilities are fragmented, or expectations are poorly aligned. By viewing the organisation as a set of interconnected processes, these issues become visible and discussable.

Therefore, the process approach provides a practical foundation for quality management. It helps organisations describe how work should be performed, clarify responsibilities across functional boundaries, and create a shared understanding of how value is created. In the following sections, this perspective is developed step by step, starting with the idea of main processes that cut across the organisation.

6.2 What do we mean by a process?

The term process is widely used in organisations, but not always with a shared meaning. It is useful to be explicit about what is meant by a process before applying the concept more broadly. In formal terms, ISO standards define a process as a set of interrelated or interacting activities that use inputs to deliver an intended result (ISO 9000).[1]

This definition is deliberately general, but it captures three essential elements: inputs, a sequence of activities, and outputs. A process exists to produce an outcome, and that outcome is intended to create value for a customer or another interested party.

This basic logic is illustrated in Figure 6.1, which shows a simple model of a process. Inputs may consist of e.g. materials or information. Through a sequence of interconnected activities, these inputs are transformed into outputs, such as products or services.

DOI: 10.4324/9781003671749-6

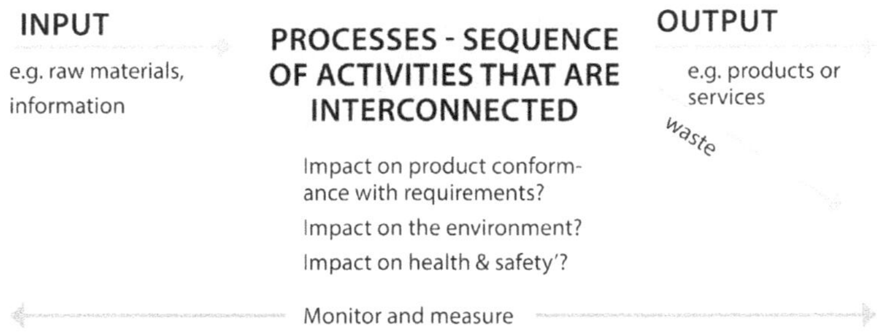

Figure 6.1 A process as a system of interrelated activities transforming inputs into outputs.

Monitoring and measurement can take place throughout the process, from input to output, providing a basis for control and improvement. A variety of aspects are considered when measuring a process, for example, to see how effective it is, whether there is a loss in the process, and how efficiently it delivers products according to expectations. It is also possible to measure the impact of the process on the environment—and on the health and safety of employees and others—and whether the process has other effects, for example, on society. It should also be kept in mind that waste or loss may occur in the process, when information is lost, or raw material deteriorates.

A simple example illustrates this logic. Consider a small hot dog stand. Most readers will have been customers in such an operation. The inputs include raw materials (sausages, buns, condiments), equipment, and the customer's order. The process itself consists of a sequence of activities, listening to the customer's request, preparing the hot dog according to that request, serving a beverage if ordered, and handling payment. The output is a correctly prepared hot dog, delivered quickly and as requested.

Although the operation is simple, it clearly functions as a process. Time is a critical factor; the faster the clerk can serve one customer, the more customers can be served in a

Figure 6.2 A hot dog stand—a simple operation?

given period. At the same time, other considerations matter. The quality of the product must meet expectations, food safety must be ensured, and the workload must not compromise the health and safety of the worker. All these aspects can be influenced by how the process is designed and carried out.

This shows that even a very simple organisation can benefit from being understood as a process. The process can be described and thus represent a shared understanding of how work is intended to be carried out. The process can be monitored and improved. Bottlenecks can be identified, unnecessary steps removed, and working conditions adjusted. The same logic applies to more complex organisations, even though the processes involved may be far more elaborate.

Understanding what is meant by a process allows the organisation to move from abstract descriptions of work to a structured view of how value is created, starting with the identification of main processes and organisational purpose.

6.3 Main processes and organisational purpose

The next step is to identify which processes are fundamental to the organisation. In this book, main processes are used to describe the organisation's core product- or service-delivery processes—those processes through which the organisation fulfils its purpose.

When mapping the main processes, it is useful to step back and look at the organisation from a distance. At this level, the objective is to ask basic questions about what the organisation delivers. What are the products of the organisation or the services it provides? Who are the customers? Who are the suppliers, and where do the inputs and resources come from that are used to create value?

Sometimes, the answers to such questions are straightforward. In other cases, there may be some ambivalence. Organisations may struggle to articulate their core offerings, particularly when services are complex or customised. Working explicitly with these questions is an important part of understanding the organisational purpose and scope.

Figure 6.3 illustrates the starting point for a process-oriented description of the organisation. At the top of the figure is strategic planning, which defines the overall direction and priorities. Below this is the level of main processes, where the organisation's core offerings are identified and described. Mapping the main processes requires answering

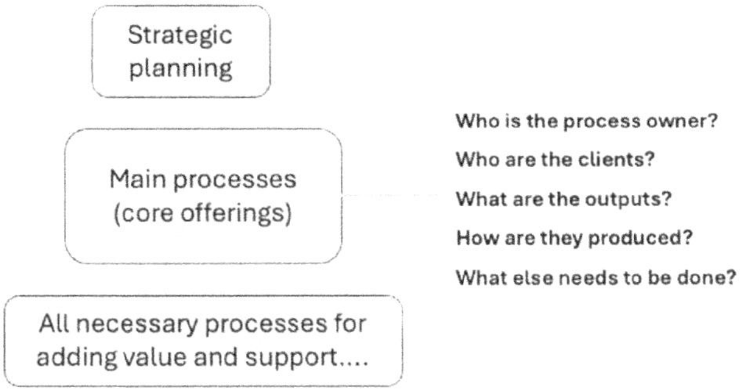

Figure 6.3 Main processes and what is needed to map them.

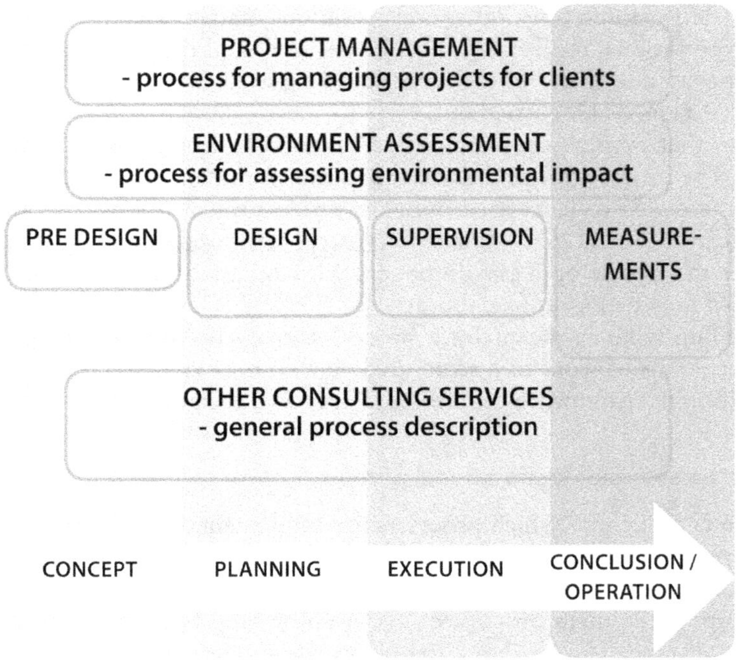

Figure 6.4 Main processes of engineering consultancy.

a small number of fundamental questions such as Who owns the process? Who are the clients? What products or services are delivered? How are they created? What additional activities are needed to support this work?

Defining the main processes shifts attention away from departments and internal structures and towards collaborative, end-to-end efforts. Most products and services are created through coordinated work across organisations. Main processes cut across organisational boundaries and make collaboration between organisational units visible.

The same underlying process logic can often be used to deliver different products and services. In an engineering consultancy, for example, pre-design and detailed design may be offered as separate services. Although the outputs differ, they may follow the same fundamental project-level process. Recognising this helps organisations avoid unnecessary fragmentation and supports learning and improvement.

Figure 6.4 illustrates the main processes of an engineering consultancy firm. The figure does not describe departments, but the principal services offered to clients, expressed in a project-oriented context. Each main process represents a distinct offering, even though the detailed execution may involve many of the same roles, competencies, and activities.

Starting from the main processes ensures that process mapping is grounded in the organisational purpose. It also provides a natural transition to the value chain perspective and the distinction between value-adding and support processes.

6.4 From main processes to the value chain

The next step is to understand how these processes contribute to value creation. For this purpose, we return to the concept of the value chain, which provides a useful conceptual

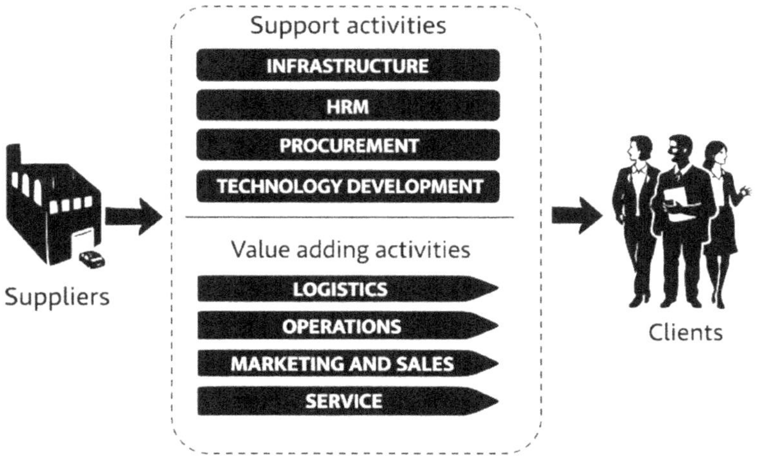

Figure 6.5 The value chain, value-adding activities, and support activities.

model. The value chain, originally introduced by Michael Porter (1985),[2] distinguishes between activities that directly create value for customers and those that support this value creation.

In the value chain logic, value-adding activities are those that contribute directly to the product or service for which the customer is willing to pay. Support activities do not directly add value from the customer's perspective, but they are necessary to enable value-adding activities to function effectively. This perspective helps clarify the different roles within an organisation's operations. The main processes typically consist of value-adding activities that transform inputs into products or services, but they depend on a range of support processes that provide resources, coordination, infrastructure, and control. Without these supporting activities, value creation becomes unreliable or unsustainable, even if the value-adding activities themselves are well designed.

A simple example illustrates this distinction. In the case of a hot dog stand, the value-adding part of the operation is easy to recognise, namely taking the customer's order, preparing the hot dog as requested, and serving it quickly and correctly. However, for this to work, a range of supporting activities must also be in place, such as purchasing raw materials, organising work shifts, maintaining equipment, and ensuring food safety. These activities do not add value directly from the customer's perspective, but without them, the value-adding process cannot operate.

Although a distinction is made between value-adding and support processes, this should not be interpreted as a distinction between important and unimportant activities. Optimisation and cost reduction are relevant to both. Improvements in support processes can substantially impact the effectiveness, reliability, and cost of value creation, even though these processes do not directly add value from the customer's perspective.

The value chain thus provides a structured way of understanding how different processes relate to one another and how value is created through coordinated effort. It also prepares the ground for extending the analysis beyond a single organisation. In many settings, value is created not only within one organisation, but also across several interconnected actors. This broader perspective is often referred to as a value network and will be briefly revisited later.

The value chain creates a clear conceptual link between the identification of the main processes and a more detailed discussion of value-adding and support processes in the following sections.

6.5 Value-adding processes

Value-adding processes are sometimes referred to as primary activities in the literature. They typically include inbound logistics, operations, outbound logistics, marketing and sales, and services. The original purpose of the value chain model was to disaggregate a company into its strategically relevant activities to identify sources of competitive advantage.

What constitutes "value-adding" depends on the type of organisation. In a simple operation, such as a hot dog stand, value-adding activities are relatively easy to identify. Inbound logistics consist of purchasing and receiving raw materials. Operations involve preparing and serving the hot dog. Outbound logistics are minimal, as customers receive the product directly at the stand, and marketing and sales may be limited to location and signage. Even in such a simple setting, the value chain helps to structure an understanding of how value is created.

In more complex organisations, the model becomes even more useful as it helps guide the analysis towards the core of operations, where value is generated. It prevents attention from being absorbed by the organisational structure or internal routines.

Our engineering consultancy provides a useful example. Figure 6.4 illustrates how such an organisation can define its main processes. In this case, the consultancy mirrors its activities in the reality of its typical client. Clients are often engaged in projects that they are preparing or executing. A typical life cycle of a construction project can be described in broad phases such as concept, planning, execution, and conclusion or start of operation. The engineering consultancy has defined its offerings in relation to this lifecycle (see Figure 6.4). Project management services may span the entire project lifecycle. Environmental assessments may also extend across several phases, reflecting their role in supporting decision-making throughout the project. Pre-design activities typically take place during the concept phase, design during the planning phase, and construction supervision during the execution phase, which is mandatory in some jurisdictions for public clients. In addition, the consultancy may offer other specialised services that are not directly tied to a single project life cycle.

The project perspective is important for two reasons. First, it reflects the reality of most of the consultancy's clients. Second, it reflects how the consultancy itself creates value. An assignment to design a house, bridge, or tunnel is a project. The same applies to environmental assessments, concept studies, and other consulting services. These assignments have a defined beginning and end and deliver outputs with specific characteristics. Projects are the fundamental units through which work is performed and value is created, and they form the basis on which services are delivered and invoiced.

Figure 6.6 illustrates this logic and provides an overview of a general project management process that applies to different services. The term *specific processes* is used to represent the different types of services or "products" offered by the consultancy. These correspond to the main processes discussed earlier. In some cases, such processes are well established and standardised; in others, they must be designed or adapted to fit the requirements of a particular assignment. At the same time, the figure shows that there is a general way of executing projects in the organisation, regardless of their specific nature. Each project is received and formally defined. A project plan is developed, a contract with the client

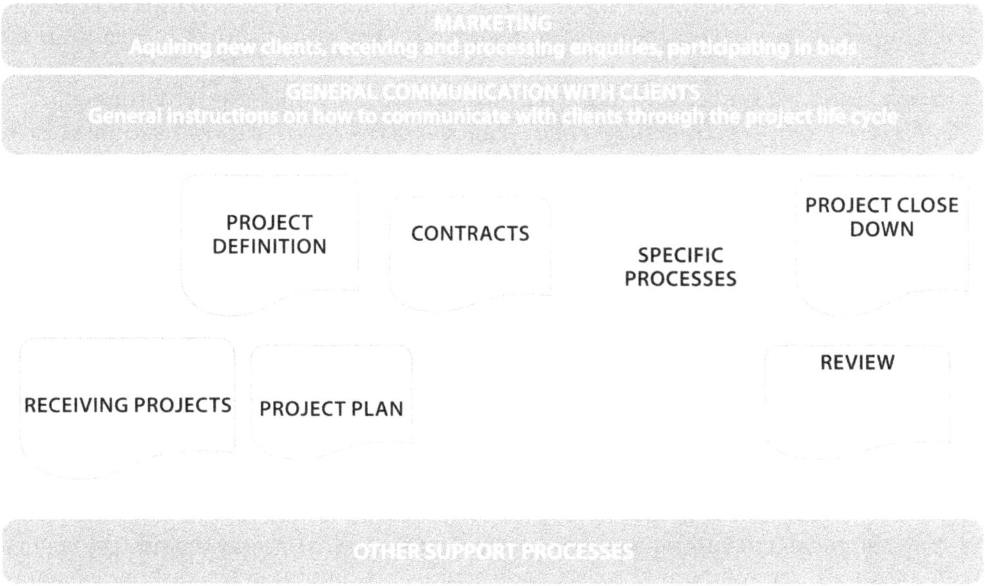

Figure 6.6 Project management process for engineering consultancy.

is formulated and signed, and specific work is carried out. Before delivery, the output is reviewed internally. Finally, the project is formally closed once delivery has taken place. These activities typically follow a largely linear pattern. In parallel, there is ongoing communication with clients and continuous marketing activities, such as acquiring new clients, responding to enquiries, and participating in bids. In project-based work, the quality of relationships between clients, consultants, contractors, and authorities can significantly influence coordination, transparency, and the ability to resolve conflicts constructively.

The value chain helps identify where value is created and how different activities contribute to the final outcome. It also highlights that similar underlying processes may be used to deliver different types of services. This insight is particularly important for organisations that rely heavily on projects, where learning from one project for the benefit of the next project is a central challenge.

This view provides a clear reference point for analysing the rest of the organisation's structure. It also sets the stage for examining support processes that enable and sustain value creation but do not add value directly from the customer's perspective.

6.6 Support processes

Support processes are essential for reliable and sustainable value creation. Therefore, support activities are enabling processes. They do not create value directly for the customer, but they shape the conditions under which value-adding activities can be performed effectively, consistently, and over time. Typical categories of support processes include:

- *Organisational governance and infrastructure.* These processes provide the structural and institutional backbone for the organisation. They include governance structures and decision-making arrangements, strategic planning and policy setting, financial management and budgeting, legal compliance, risk management, and internal controls, such as performance measurement and reporting. They also ensure that organisational activities comply with regulatory and societal expectations, thereby safeguarding the organisation's legitimacy and long-term acceptance in its operating environment.

- *People and competence management.* This concerns how an organisation attracts, develops, organises, and retains people. It may include recruitment and onboarding, skill and professional competence development, career paths, incentives, rewards, recognition, work environment and well-being, and organisational culture development. In some organisations, this is a formal HR function; in others, it is handled informally by line managers or founders.
- *Knowledge, methods, and improvement capabilities.* This category includes the development and refinement of methods and ways of working, learning from experience, and the use of data, analysis, and evidence in decision-making. It also includes innovations in services, processes, or delivery models, as well as the adoption and adaptation of external knowledge. In a manufacturing firm, this may involve product and process design; in a public agency, it may involve policy instruments or digital services.
- *Resource acquisition and external inputs.* These activities cover how an organisation secures inputs that it cannot or should not produce itself. This may include the acquisition of goods, services, or expertise; contracting and supplier relationships; outsourcing and partnerships; and access to shared services, platforms, or infrastructure. In some organisations, this is handled by a professional procurement unit; in others, it is managed ad hoc by project managers or team leads.

In the hot dog stand discussed earlier, support processes are mainly concerned with keeping daily operations running smoothly rather than developing new products or services. They include basic planning and coordination of opening hours and shift schedules, handling cash flow and supplies, and ensuring that staff are available and trained to serve customers consistently throughout the day. Even in such a small operation, these support processes create the stability that allows the value-adding activity—selling hot dogs to customers—to take place reliably from morning to midnight.

In a project-oriented engineering consultancy serving construction clients, support activities provide the continuity and coherence that make effective project delivery possible. They include organisational governance and financial steering, such as strategic positioning, pricing principles, contract and risk management, and control of project profitability and cash flow. They also include people and competence management, covering recruitment, workload and capacity planning across projects, professional development, mentoring and performance management.

Another core support activity is the development and maintenance of design guidelines, interpretation of codes and regulations, internal reviews, and quality assurance practices that safeguard professional credibility and manage technical risks. In addition, consultancies rely on shared project management frameworks, models, templates, tools, and support functions that enable consistent planning, control, and reporting across projects. Finally, business development and client relationship support underpin the entire value chain by structuring how work is identified, proposals are prepared, and long-term client relationships are maintained, even though these activities are often carried out by the same professionals who later deliver the projects.

While support processes are important, confusing them with value-adding processes can obscure where value is created and complicate responsibility and prioritisation.

Understanding the relationship between value-adding and support processes helps clarify the roles, interfaces, and dependencies within the organisation. It also prepares the ground for representing the processes at different levels of abstraction.

6.7 Processes, procedures, and their representation

Before moving into practical process mapping, it is important to clarify the distinction between a process and a procedure. Although these terms are sometimes used interchangeably, they serve different purposes in a quality management system.

A process describes what happens when inputs are transformed into outputs to create value. Processes cut across organisational boundaries and focus on work, responsibility, and outcomes. In this chapter, processes have been discussed at different levels, from main processes and value-adding processes to support processes, all of which together describe how the organisation operates.

In contrast, a procedure is a documented and agreed-upon method for carrying out a part of a process or activity. Procedures describe how work should be performed in practice in a consistent, understandable, and suitable manner for the organisation as a whole or for a specific function. A procedure typically states its purpose and scope, identifies responsibilities, and describes key steps involved. Its role is not to capture every technical detail but to provide sufficient guidance to ensure that work is conducted in a controlled and repeatable manner. In many cases, procedures are supported by more detailed work instructions, checklists, or templates, which provide step-by-step guidance for specific tasks.

Procedures can be presented in different forms depending on their purpose and users. They may be written as text, structured as tables, illustrated with diagrams, or presented as flowcharts, or as a combination of these. A flowchart is therefore not a process in itself, but a way of representing how a procedure is carried out in practice. It visualises the steps and decisions involved in carrying out part of a process. Flowcharts are particularly useful when the sequence of activities, decision points, or handovers must be made explicit.

This distinction can be illustrated with a simple example of a hot dog stand. A process-oriented view shows the main value-adding process of serving customers, supported by activities such as staffing, purchasing, and basic administration. This high-level picture helps clarify how value is created and how different activities are related to each other. A procedure, on the other hand, describes how a specific part of this process is carried out in practice—for example, how a customer order is taken, prepared, paid for, and delivered. A flowchart can effectively present such a procedure, showing the steps involved and the choices that may arise.

The distinction between processes and procedures helps ensure that process mapping and documentation efforts are purposeful and proportionate. This distinction is further discussed by explaining how processes are translated into practical procedures and how different forms of flowcharts can be used in this context.

6.8 Mapping processes in practice

Process mapping is a means of improving understanding. Organisations map processes to clarify how work is carried out, how activities are connected, and where responsibilities lie. A well-designed process map helps to make interfaces, dependencies, and weaknesses visible.

Figure 6.7 A simple checklist in a hot dog stand.

In practice, it is neither necessary nor desirable to develop procedures for every aspect of a process. Procedures are typically most useful where work is repetitive, coordination between roles or units is required, risks are significant, or consistency is particularly important. In many organisations, this means that procedures focus on selected parts of value-adding processes and critical interfaces between value-adding and support activities.

In the hot dog stand example, the person serving customers has a streamlined routine and does not require a flowchart for the basic activity of serving a hot dog, at least after elementary on-the-job training. However, some aspects of running the stand are less routine and may benefit from a formalised procedure represented by a flowchart or a checklist. Examples include receiving bread and meat from suppliers and verifying that deliveries comply with purchase orders and quality requirements or carrying out regular cleaning of equipment, activities that may even be required by local health authorities.

In this book, process mapping is promoted as a key element in the practical build-up of a quality management system (QMS). The suggested approach is to use the quality pyramid as a guiding model and start from the top. This means identifying the main processes first and then, where needed, breaking them down into more detailed sub-processes. This top-down approach is useful for developing a shared understanding, aligning perspectives, and establishing a common reference point.

More technical descriptions of specific activities require a different approach, starting with how work is performed. Such a bottom-up approach is often used when concrete problems are addressed and typically relies on the experience of those who carry out the work.

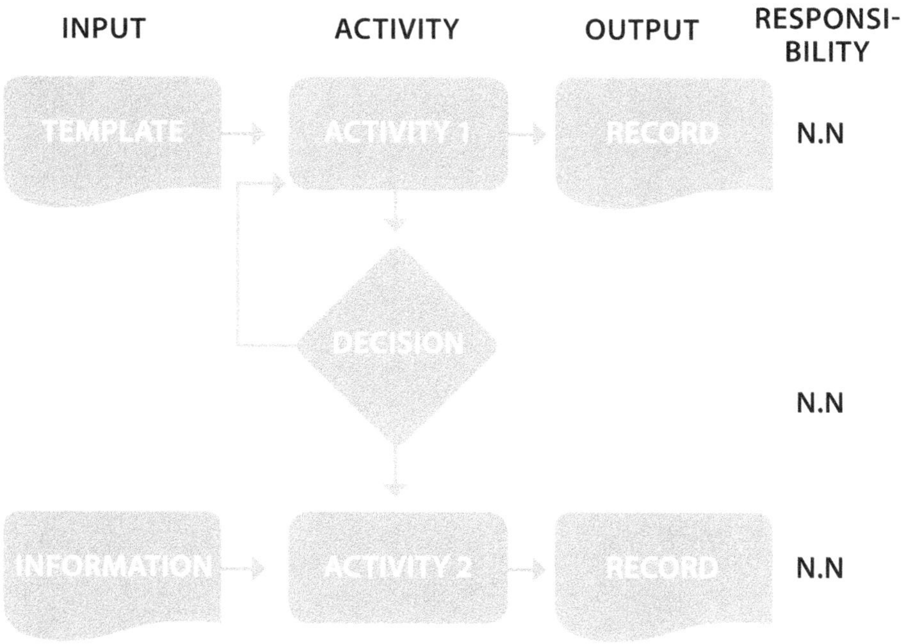

Figure 6.8 A simple flowchart showing input, activity, output, and responsibility.

Flowcharts play a central role in describing procedures in more detail. Equally important is who is involved in such efforts. Involving practitioners helps ensure that descriptions are accurate and builds a shared understanding and ownership. It also helps surface informal practices and workarounds that are often invisible in formal descriptions but may have significant implications for quality and performance.

A simple sequential flowchart can be used to describe the main steps in a procedure and the order in which they are performed. An example of such a flowchart is presented in Figure 6.8.

Figure 6.8 illustrates how a procedure can be represented using a flowchart. The figure shows how different types of input, such as a predefined template or case-specific information, enter the procedure and are transformed through a sequence of activities into documented outputs, here referred to as records. The decision point highlights that procedures often include choices that influence the subsequent flow of work rather than a single linear sequence of steps. By explicitly linking inputs, activities, decisions, outputs, and responsibilities, the flowchart clarifies what happens, based on which information, who is responsible at each step, and with which result. This type of flowchart is useful when the objective is to understand the basic logic of a procedure without prescribing unnecessary technical details.

A swim lane flowchart represents a procedure in which activities are organised according to responsibility. The diagram is divided into horizontal or vertical lanes, where each lane represents a role, function, or organisational unit. Activities are placed in the lane of the actor responsible for carrying them out, while arrows show the sequence of work and handovers between roles. This makes swim lane flowcharts particularly useful for procedures that involve coordination across several departments.

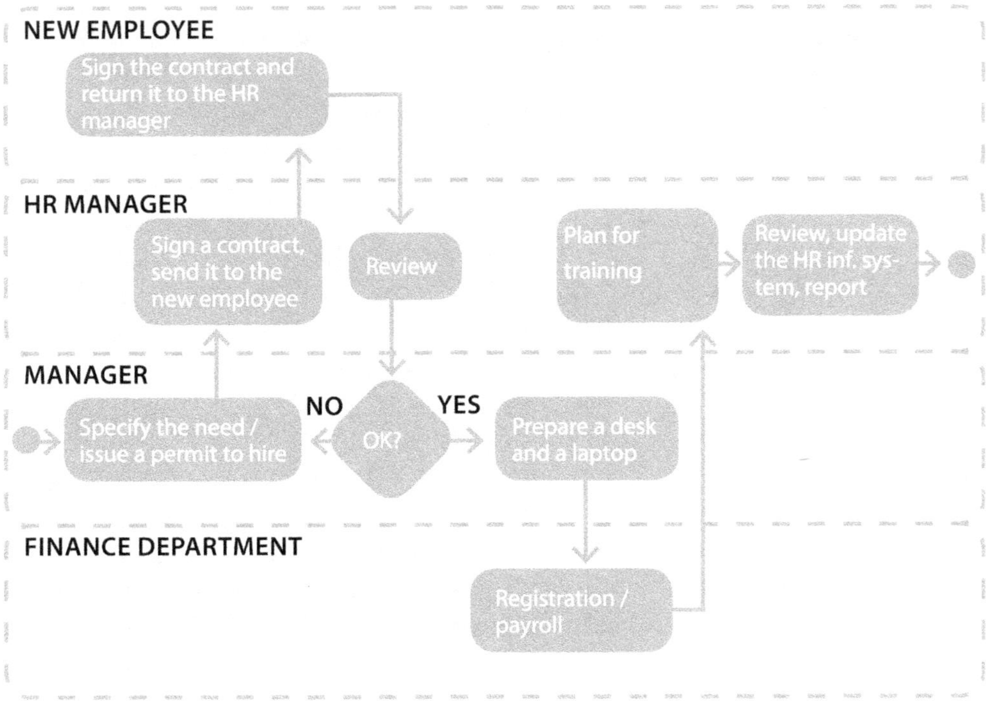

Figure 6.9 A swim lane flowchart showing a simple process for hiring new employees.

The logic of this flowchart is illustrated in Figure 6.9 represents the procedure for hiring and onboarding new employees. The lanes correspond to the manager, HR manager, new employee, and finance department. The procedure begins with the manager specifying the need and issuing the permission to hire. A review and decision step determines whether the process continues. If approved, activities are triggered in several lanes. HR prepares and exchanges the contract with the new employee, practical arrangements such as preparing a desk and laptop are made, payroll registration is handled by finance, training is planned, and HR systems are updated. The figure clearly shows how the procedure cuts across organisational boundaries and how responsibilities are distributed while still forming a coherent and coordinated sequence of actions. Such representations are sometimes integrated into digital workflow systems and enterprise platforms, enabling coordination and transparency across organisational boundaries. In project-oriented organisations, such procedures often play an important role in ensuring that information, decisions, and handovers are communicated clearly between project participants.

A swim lane flowchart should be distinguished from more formal modelling approaches, such as the Business Process Model and Notation (BPMN). BPMN is designed as a standardised way of modelling business processes that can be understood and used by both business and technical stakeholders. It uses a defined set of symbols for events, activities, gateways, and flows and provides a precise language for representing complex process logic. While BPMN can be very powerful, swim lane flowcharts are often sufficient—and more accessible—when the purpose is to describe procedures, clarify responsibilities, and support communication and learning within an organisation.

Process mapping should not be considered a one-off exercise. Organisations change, and processes evolve as products, technologies, and requirements develop. Process maps and procedures should be treated as living descriptions of how work is performed. Structured processes and well-designed procedures provide stability in daily operations while allowing adaptation as requirements, technologies, and environments change. When used in this manner, they provide a natural foundation for monitoring, measurement, and improvement.

6.9 From value chains to value networks

The process and procedure perspectives discussed in this chapter are most naturally developed using the value chain as a starting point. The value chain provides a powerful and intuitive way to understand how organisations create value through a sequence of interconnected activities. For many purposes, this linear representation is sufficient and highly useful for analysis. However, in many real-world settings, value creation cannot be fully understood as a simple chain that runs from input to output within a single organisation.

Increasingly, organisations operate within value networks, where multiple actors interact in non-linear ways to create and deliver value. In such settings, value is produced through the exchange of resources, information, and services across organisational boundaries, often involving customers, suppliers, partners, and platforms. Relationships, coordination mechanisms, and shared rules become central to coordinating work across organisational boundaries. In this context, coordination across organisational boundaries depends on clearly defined procedures and well-managed interfaces between organisations. Value creation increasingly includes broader expectations, such as sustainability, safety, and long-term responsibility, extending beyond immediate customer satisfaction.

This network perspective is closely related to the systems view of organisations. Rather than treating the organisation as a closed entity, it highlights its character as an open system that both influences and is influenced by its environment. This has important implications from a quality management perspective. Control, coordination, and improvement extend beyond internal processes and increasingly depend on how an organisation interacts with others.

A familiar example is Uber, which can be understood not as a traditional service provider with a linear value chain, but as a digital ecosystem. Uber creates value by enabling interactions between drivers and passengers, supported by a digital platform, pricing mechanisms, and reputation systems. Value is co-created by multiple actors, and quality depends not only on internal processes, but also on behaviour, incentives, and feedback loops within the network.

The concept of value networks does not replace the value chain but complements it. The value chain remains a useful starting point for understanding internal operations and structuring process mapping. The value network perspective adds an additional layer, helping to explain the complexity, interdependence, and shared responsibility in contemporary organisations.

Recognising when a value chain perspective is sufficient and when a value network perspective is needed is useful. It allows organisations to choose appropriate ways of analysing, coordinating, and improving their operations in increasingly interconnected contexts.

Case: The engineering consultancy as a project-based organisation

The engineering consultancy described earlier delivers its value almost exclusively through projects. Each assignment is time-limited, often unique in technical scope, and is carried out in close interaction with clients, contractors, and authorities. Unlike a manufacturing company, it does not produce a steady stream of identical outputs. Its product consists of designs, analyses, calculations, and professional recommendations—intangible results whose quality may only become fully visible when implemented. Therefore, planning accuracy, coordination, and professional judgement are central elements in ensuring the quality of services.

For such an organisation, the process perspective requires careful reflection. If every project is different, how can consistency be achieved? If outputs are intangible, how are requirements clarified and verified? If projects are temporary, how does experience accumulate rather than disappear when the project closes? The reader is invited to reconsider the consultancy's value chain and process structure considering these challenges. Where is quality secured in the organisation's processes? Where is the learning anchored? What mechanisms ensure that each completed project strengthens the next?

Reflection points—Chapter 6

1. Process or hierarchy?
   In many organisations, authority follows the organisational chart, whereas value creation follows processes that cut across functions. Consider a public agency that you are familiar with. Are responsibilities aligned with value-creating processes or primarily with departments? What happens when process ownership cuts across formal authority? Does it improve coordination or create ambiguity and conflict? Reflect on the trade-off between hierarchical clarity and cross-functional effectiveness.
2. Standardisation or professional judgement
   In a project-based engineering consultancy or IT services firm, every assignment is unique; however, the procedures aim to ensure consistency. How far should procedures prescribe behaviour before they begin to constrain professional competence? Where is the boundary between necessary control and excessive formalisation? Discuss the trade-off between reliability and flexibility.
3. Value creation or compliance
   Support processes often focus on governance, documentation, and regulatory compliance. Consider a large organisation that you are familiar with. Do support processes primarily enable value creation, or do they risk becoming ends in themselves? Discuss whether and how support mechanisms can dilute operational capabilities.
4. Stability or adaptability
   Well-designed processes provide structure and predictability, whereas turbulent environments demand responsiveness. In sectors such as digital services, how does an organisation balance stable procedures with the need for rapid change? Discuss whether process stability functions as a foundation for resilience or as a barrier to innovation.

5. Linear chain or value network
   Traditional value chain thinking assumes a relatively contained system, but many organisations, such as platform-based mobility services or logistics networks, operate in ecosystems. Think about the organisation where you work: does it manage quality primarily within its own boundaries or across partnerships and digital platforms? Where are interfaces with suppliers, users and regulators explicitly governed? Analyse the risks that emerge when quality responsibilities are distributed across actors.
6. Customer value or societal value
   When processes are designed primarily around customer requirements, broader impacts, such as environmental footprint or social responsibility, may remain secondary. Consider a sector that you are familiar with. Are sustainability and long-term societal considerations embedded in core processes or treated as external constraints? Where in the value chain would such considerations need to be anchored to influence the decisions?

Notes

1 International Organization for Standardization. (2015). ISO 9000:2015 Quality management systems—Fundamentals and vocabulary (ISO Standard No. 9000:2015).
2 Porter, M.E. (1985). *Competitive Advantage: Creating and Sustaining Superior Performance.* New York: Free Press.

7 Organising for quality
Structure, roles, and information

7.1 The organisation as a social system

Organisations are often described as systems, but not in the sense of mechanical or technical systems that can be fully specified, optimised, and controlled through formal design alone. An organisation is a social system shaped by people, relationships, shared meanings, informal practices, and patterns of interaction. Formal structures, procedures, and information systems are important; however, they cannot operate in isolation from human behaviour and interpretation.

This perspective builds directly on the systems view introduced earlier in this book. Organisations were described as open systems, interacting continuously with their environment through suppliers, customers, regulators, and other stakeholders. Examples such as hotels, fast-food restaurants, and construction projects illustrate how quality outcomes emerge from the interaction between design decisions, internal activities, and external influences rather than from isolated functions or procedures.

This distinction is fundamental. A quality management system cannot be successfully operated as a purely technical arrangement of procedures, documents and controls. Its effectiveness depends on how people understand, use, trust, and integrate it into their daily work. Therefore, leadership, communication, professional norms, and organisational culture are decisive factors.

This view reflects systems thinking as it has been developed in organisation theory and management, including the idea of the learning organisation as articulated by Peter Senge (1990)[1]. Organisations learn through feedback, continuous learning, and collaboration. They adapt their behaviour over time, and their performance improves through shared understanding and collective reflection on what works and what does not.

The idea of a learning organisation is particularly useful in the context of quality management. A QMS both shapes and is shaped by the organisation in which it operates. It influences how people communicate, how responsibility is understood, how problems are addressed, and how improvements are pursued. It is also constrained by the existing norms, values, and power relations. Employee participation, interpretation, and professional judgement are central to how the system functions.

Quality management must be connected to an organisation's broader purpose and long-term direction. When quality is understood not only as compliance with requirements but also as a contribution to sustainable value creation, it becomes a structural driver of responsible behaviour and long-term performance.

The systems view provides the foundation for the remainder of this chapter. It explains why organisational culture is decisive for quality management, why roles and governance

DOI: 10.4324/9781003671749-7

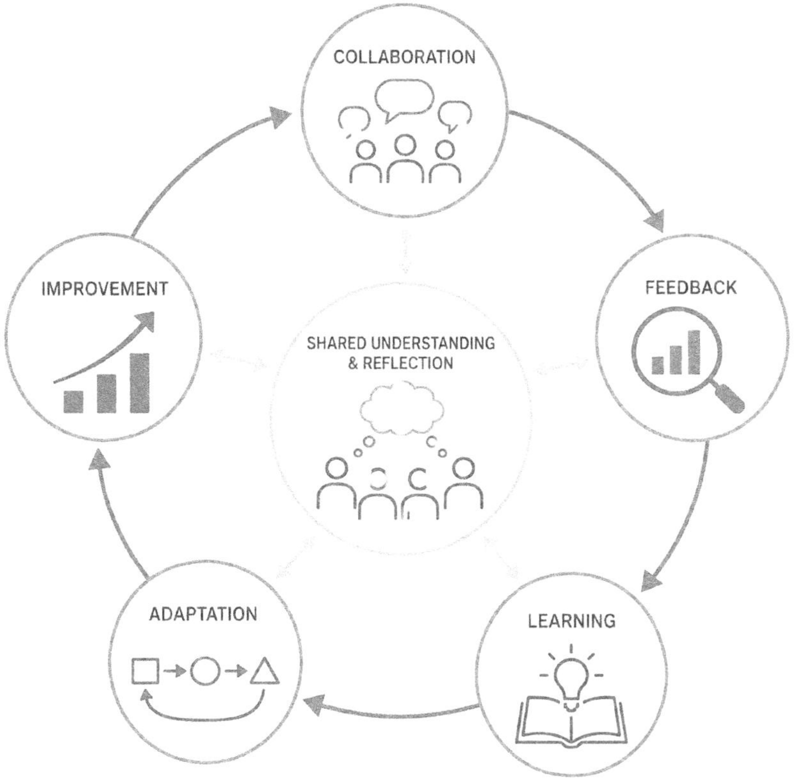

Figure 7.1 The learning organisation.

structures must be clearly understood and visible, and why communication and information flows are critical.

7.2 Organisational culture and quality management

Organisational culture plays a decisive role in quality management efforts. It shapes how people understand their work, interpret rules and procedures, react to problems, and how willing they are to change established ways of working. For many organisations, the success or failure of a QMS depends on how well it aligns with existing cultural patterns rather than the technical design of the system.

This is particularly important because introducing or developing a QMS almost always involves changes. Even for a well-structured and professional organisation, a formal QMS may introduce new expectations regarding transparency, documentation, feedback, and responsibility. In organisations with little tradition of formal procedures, these changes can be experienced as intrusive or unnecessary. Implementing a formal QMS should therefore be understood as a change effort that interacts directly with organisational culture.

A common mistake is to move too quickly to define desired procedures or future ways of working without first understanding the present situation. It is usually essential to begin with a status assessment of the current situation. This includes technical aspects, such as existing procedures, documents, and information systems, but also cultural aspects, including prevailing attitudes towards quality, responsibility, and improvement.

Organisational culture is an abstract concept, and for many readers it may initially seem difficult to grasp. It does not consist of formal rules alone, nor can it be fully captured in organisational charts or policy documents. Culture is reflected in everyday behaviour: how decisions are made, how problems are discussed, how mistakes are handled, and how people talk about their organisation. It is often expressed indirectly through routines, habits, informal norms, and shared assumptions.

When working with quality management, it is therefore useful to ask practical questions such as the following:

- How are problems typically handled in the organisation?
- Are mistakes discussed openly or hidden?
- Do employees feel responsible for the quality of their work?
- How do managers react when deviations are reported?
- Is improvement seen as a shared responsibility or as someone else's task?

These questions help to reveal the cultural patterns that strongly influence how a QMS functions in practice. An organisation where employees already take pride in their work and discuss problems openly will typically find it easier to formalise practices through a QMS. In contrast, an organisation where responsibility is unclear or where reporting problems is perceived as risky may struggle, regardless of how well the system is designed on paper.

Therefore, understanding organisational culture at an early stage is about developing a realistic picture of how the organisation works. This understanding provides an essential foundation for decisions regarding how quality management should be introduced, communicated, and developed.

Figure 7.2 How are problems typically handled?

In the following sections, this general discussion of organisational culture is made more concrete. First, a practical framework, the culture web, is introduced to analyse and discuss organisational culture in a structured manner. This is followed by a more focused discussion of quality culture, which highlights attitudes and behaviours that are particularly relevant in the context of quality management.

7.3 Understanding organisational culture

Organisational culture is often experienced as something intangible and difficult to describe; however, it has a strong influence on how organisations function in practice. Professional managers need a structured way of discussing culture and making its main characteristics visible. One such tool is the culture web (Johnson et al. 2008)[2], which provides a practical framework for analysing organisational culture in a systematic way.

The culture web focuses on observable elements of organisational life that shape how people behave and how decisions are made. By explicitly discussing these elements, it becomes easier to understand why certain practices are accepted, why others are resisted, and how quality-related initiatives are likely to be received. The culture web consists of six interrelated elements:

Symbols include words, images, artefacts, and actions that carry meaning within an organisation. These may be formal, such as titles, slogans, office layouts, or documented values, but they are often informal and implicit. Examples include recurring expressions, jokes, or symbolic actions by managers. Symbols send strong signals about what is valued and taken seriously. Symbols may reveal whether quality is

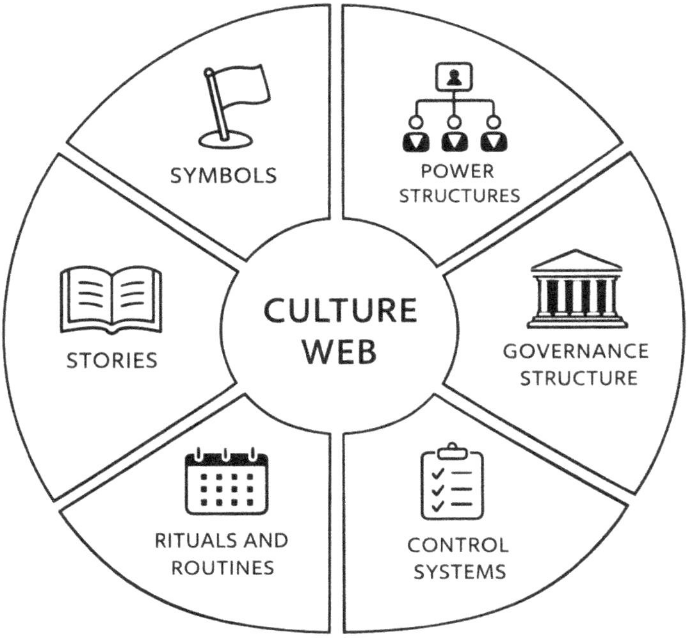

Figure 7.3 Culture web—a tool to assess the culture of organisations.

regarded as central to the organisation's identity or merely as a formal requirement. For example, a respected utility company used a hand hatchet to symbolise change in the will to improve, cut down costs, and increase efficiency. When the tool was handed over to a departmental manager within the organisation, it was a clear message to that person about what he or she had to do.

Power structures describe where real influence resides within an organisation. Formal authority, as shown in organisational charts, does not always correspond to actual power. Some individuals or groups may have a disproportionate influence because of their expertise, experience, personal relationships, or historical position. Understanding power structures is important for quality management because they affect who can initiate change, whose opinions carry weight, and how conflicts related to quality issues are resolved.

The **governance structure** reflects how responsibilities and roles are formally organised. It shows how departments are connected, how decisions are escalated, and where accountability lies. From a quality management perspective, governance structures influence the clarity of quality-related responsibilities and the consistency of practices across the organisation. A highly hierarchical structure may signal strong central control, whereas a flatter structure may indicate greater delegation and employee involvement.

Control systems include formal and informal mechanisms used to monitor performance and behaviour. These may involve measurements, key performance indicators (KPIs) such as defect rates, delivery reliability, customer satisfaction, or rework levels, reporting routines, audits, incentives, and reviews. What is measured, how results are communicated, and how deviations are handled provide strong indications of organisational priorities. Control systems are closely linked to quality management because they influence attention, behaviour, and learning. Some organisations emphasise strict follow-up and detailed control, while others rely more on professional judgement and looser forms of oversight.

Rituals and routines refer to habitual ways of working and interacting within an organisation. These include meetings, training practices, onboarding of new employees, performance reviews, and so on. This can also refer to informal events such as hosting an annual Christmas ball for the children of employees, eating breakfast together every Friday morning, or organising a shared social event when the project is delivered to the client. They reveal what is considered normal and acceptable behaviour. In quality management, routines such as regular reviews, improvement meetings, and feedback sessions can support learning and continuous improvement, provided they are taken seriously.

Stories are narratives that circulate within an organisation about past events, successes, failures, and key individuals. Reputation reflects how an organisation is perceived internally and externally. Together, stories and reputation shape how employees interpret expectations and evaluate which behaviours are rewarded or discouraged. Stories about how problems were handled in the past can strongly influence people's responses to quality issues in the present.

These six elements form a coherent picture. They reinforce each other and are relatively stable over time. The culture web is useful as a diagnostic and discussion tool. It helps managers and quality professionals identify cultural patterns that may support or hinder quality-related initiatives in an organisation. This understanding can be used as a basis for informed decisions regarding the introduction, communication, and development of quality management.

In the next section, this broad understanding of organisational culture is narrowed to a more specific concept, quality culture. This focuses on attitudes and behaviours that are directly relevant to meeting requirements, preventing errors and improving performance.

7.4 Quality culture and its assessment

While organisational culture describes the broader patterns of values, routines, and power relations within an organisation, quality culture focuses specifically on attitudes and behaviours related to quality. It concerns how people think about meeting requirements, preventing errors, learning from experience, and taking responsibility for their work. This distinction is useful because it allows attention to be directed towards cultural factors that have a direct impact on quality outcomes.

Quality culture influences the perception and practical application of a quality management system. It affects whether procedures are followed because they are seen as meaningful, whether deviations are reported openly, and whether improvement activities are treated as opportunities for learning or as unnecessary disturbances. Two organisations with similar formal QMSs may therefore perform very differently, largely because of differences in their quality culture.

A simple and widely used way of describing different orientations towards quality culture is provided by Cameron and Sine (1999),[3] who distinguish between four broad categories:

- No focus on quality, where quality-related issues receive little attention and problems are often ignored or dealt with informally.
- Identifying defects, where the primary concern is to ensure that defective products or services do not reach customers.
- Mistakes are avoided, where information about errors and deviations is actively used to prevent recurrence.
- Creative quality culture, where the organisation focuses on understanding customer needs and continuously improving how value is created, often aiming to exceed expectations.

This simple model is a useful first step in understanding the attitudes within the organisation, but it is also important for leading the implementation of a QMS. For example, consider an organisation in the fourth category, trying to surpass customer expectations. This kind of company is likely to be running a system to control all its processes and production, and it probably takes good care of all communication with its clients and interested parties. However, even if this type of system is not formally in place, its implementation will be an easy task.

An organisation at the other end of the scale, with an undeveloped quality culture, will need to invest considerable effort in shaping attitudes before the formal development and implementation of a QMS can begin. This typically requires education, training, and sustained leadership attention. Nevertheless, some changes may be difficult to achieve, as not all employees are equally willing or able to adapt to their established ways of working.

These categories are not strict classifications. Most organisations display elements of more than one category, and different parts of the same organisation may exhibit different quality cultures. The framework is useful for understanding why quality management initiatives unfold differently across organisations.

Understanding the quality culture is particularly important in the early stages. Before deciding how formal procedures should be designed or how extensive documentation should be, it is necessary to understand the prevailing attitudes towards responsibility, transparency, and improvement. This helps explain which elements of a QMS are likely to be readily accepted and which may require additional explanations, training, or leadership attention.

One way to assess employees' perceptions of quality management is to facilitate discussions among them. Examples of questions to start the conversation could be the following:

- Do you have a clear understanding of what quality means in your daily work?
- How visible is the organisation's emphasis on quality in its decisions, priorities, and communication?
- Do the actions and attitudes of your immediate manager demonstrate that quality is taken seriously?
- When problems or deviations occur, are they discussed openly and used as a basis for learning?
- Do you feel responsible for the quality of your own work, and do you have the means to influence it?
- How well do formal procedures reflect the way work is conducted in practice?
- Have you participated in activities related to improvement or quality development in the last 12 months?

All these questions can be the basis of a valuable and enlightening discussion regarding quality management in an organisation. To make the concept of quality culture even more concrete, it is useful to identify observable characteristics that tend to be present in organisations with a strong quality culture. Drawing on the work of Goetsch and Davis (2012)[4], such organisations often display the following features:

- Consistency between what the organisation claims to value and how it behaves.
- Active collection and use of customer feedback as input for improvement.
- Employee participation in shaping and improving work processes.
- Extensive use of teamwork, both internally and in work with clients.
- Visible commitment to quality by the senior management.
- Availability of sufficient resources to support quality management.
- Ongoing education and training, with an emphasis on preserving and developing knowledge.
- Recognition and reward systems that support quality-related behaviour.
- Understanding colleagues as internal customers.
- Suppliers are viewed as partners rather than adversaries.

These features provide a useful basis for assessing the quality culture in practice. One practical approach is to use them as the foundation for structured discussions or surveys within the organisation. Employees can be asked to reflect on the extent to which these characteristics are present in their daily work, for example, using a simple scale ranging

	Does not matter at all	Does not matter	Neutral	Matters	Matters very much
There is coherence between slogans and actions					
Customer feedback is sought and utilized					
The employees are participants					
Teamwork is common					
Senior managers are committed					
There are enough resources					
There is adequate education and training					
Promotion and salary systems take participation in the QMS into account					
Co-workers are seen as internal clients					
Suppliers are seen as business partners					

Figure 7.4 A survey to assess the quality culture of an organisation.

from "does not matter at all" to "matters very much." The purpose of such an assessment is not to produce a precise measurement, but to stimulate reflection and dialogue about strengths, weaknesses, and areas for improvement.

Conducted in a workshop setting, such an assessment can help create a shared language, reveal differences in perception between groups or organisational levels, and provide a starting point for prioritising improvement efforts. It also helps set realistic expectations regarding the pace and direction of change.

Quality culture develops over time through experience, leadership behaviour, feedback, and learning. Such development can be supported by making expectations explicit, providing information about performance, and creating arenas for reflection and improvement.

7.5 Professional autonomy, professionalism, and responsibility

Quality management ultimately takes place in daily work. Regardless of how well a QMS is designed, its effectiveness depends on how people understand their roles and responsibilities, perform their tasks, make judgements, and respond to deviations in real situations.

Responsibility lies on everyone's shoulders, but the beginning and end of the implementation and operation of a QMS in any organisation is the active participation, dedication, and follow-up of the top management. This is closely related to organisational culture, awareness among senior leaders, and their understanding of the ultimate purpose of quality management. Top managers must have basic knowledge and recognise that

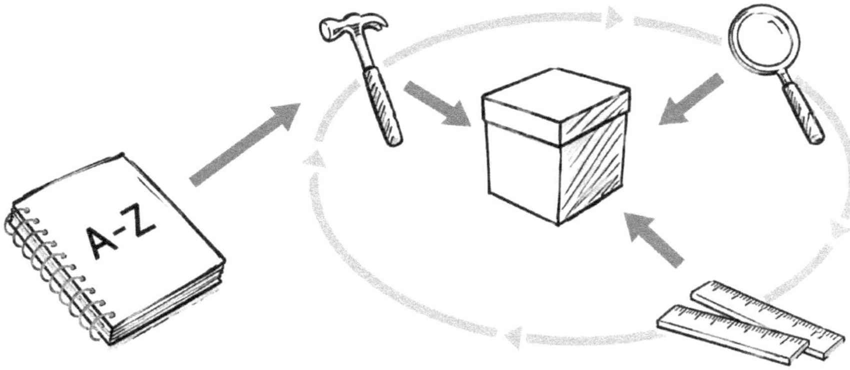

Figure 7.5 The concept of professional autonomy explained graphically.

considerable effort is required, not least from themselves. They must demonstrate interest and focus, take an active role, and act as role models in the implementation and ongoing operation of the system.

This brings attention to a central idea. An employee who is doing a certain job should be able to control what he or she is doing. Professional autonomy (sometimes referred to as self-control) refers to an organisational arrangement in which employees are responsible for monitoring and managing their own work. Rather than relying on close supervision and detailed instructions from above, quality is achieved through professional judgement, access to relevant information, and clearly understood expectations. In such arrangements, the role of management is to create conditions that enable responsible and competent behaviour.

This idea contrasts with more traditional forms of management, where work is organised through detailed orders and close supervision. While such arrangements may function in stable and repetitive environments, they are poorly suited to contemporary organisations characterised by complexity, knowledge work, and continuous change.

From a quality management perspective, professional autonomy has several advantages. Employees who understand what is expected of them and have the competence to assess their performance are often best positioned to prevent errors. They are closer to the work, are more familiar with its details, and are more likely to notice early signs of deviation. This supports both quality and efficiency while reducing the need for costly layers of management and control. The relevance and extent of professional autonomy, however, vary across different types of work. In highly standardised or safety-critical environments, formal procedures and supervision may play a stronger role, while in knowledge-intensive or project-based settings, professional autonomy becomes a central mechanism for ensuring quality.

Certain conditions must be met for professional autonomy to function in practice. First, employees must have the necessary knowledge and clear instructions on what is expected of them. They must understand the requirements, standards, and boundaries, what must be achieved, what is acceptable, and what actions should be taken when problems arise.

Figure 7.6 Daily scrum meeting—an example of professional autonomy.

Second, employees must have access to information regarding their performance. This includes feedback on outcomes, deviations, customer responses, and other relevant indicators. Without such information, it is difficult for individuals to assess whether their work meets the expectations.

Third, employees must be able to act when deviations are detected. This requires a degree of trust and delegation, as well as clear escalation paths when issues exceed an individual's authority or competence to address them.

A project-related example of this principle can be found in agile project management, particularly in the Scrum framework. A typical Scrum team consists of five to seven skilled specialists who work together as a self-organising unit. Rather than being directed by a traditional line manager, the team receives prioritised requirements from a product owner and agrees internally on how to deliver them within a short development cycle—often called a sprint—lasting two to three weeks.

Within the sprint, the team plans its own work, distributes tasks according to competence, monitors progress continuously, and resolves problems collaboratively. Responsibility for quality lies within the team, and errors are detected early through frequent reviews and reflection.

Professional autonomy is based on the clarity of expectations, transparency, and trust in professional competence. When these elements are present, quality is built into the work itself rather than being inspected afterwards.

Organisations with a strong quality culture tend to support professional autonomy by encouraging transparency, learning, and responsibility. Attempts to introduce professional autonomy in environments characterised by fear, blame, or unclear expectations are unlikely to succeed. Professional autonomy cannot be imposed through formal rules alone; it must be supported by consistent leadership behaviour and a culture that values professionalism.

Professional autonomy links organisational culture and formal structure in practice. It shows how values such as responsibility and quality are translated into everyday behaviour and highlights the importance of organisational arrangements that support professional judgement. In the following sections, this perspective is developed further by examining how governance structures, roles, and organisational design can enable professional autonomy and make responsibility clear and visible.

7.6 Organising for quality: governance structure and roles

Governance structures provide a formal framework within which quality management is coordinated, decisions are made, and accountability is exercised.

The governance structure describes how roles, responsibilities, and authority are formally organised within an organisation. It shows how different functions and units are connected, how information flows, and where decisions are made. In most organisations, this structure is represented through an organisational chart (often referred to as an organogram). While organisational charts never tell the full story of how work is actually done, they play an important role in clarifying expectations and making responsibilities explicit.

In quality management, governance structures are important for several reasons. First, they determine who has the authority to define policies, approve procedures, and allocate resources for quality-related tasks. Second, they influence the consistency of quality practices across departments and projects. Third, they signal whether quality is treated as a core management responsibility or peripheral support activity.

In a traditional functional organisation, work is divided into departments such as production, design, marketing, finance, and human resources. These departments are often

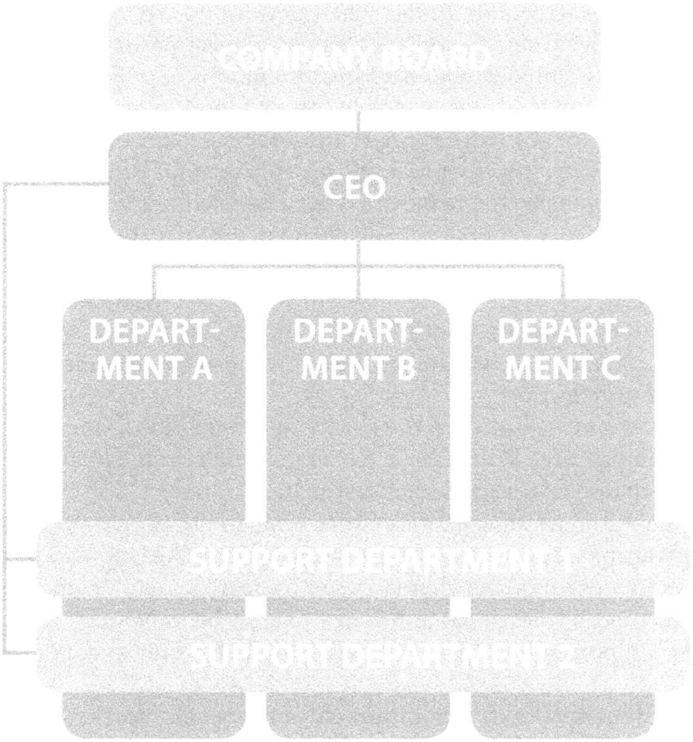

Figure 7.7 Typical organisational chart showing functional and support departments.

the location of specific professional expertise. Many organisations also show functions, such as IT, HR, or finance, that cut across functional boundaries. These functions enable coordination, consistency, and learning across the organisation.

When quality management is introduced in such a setting, an important question arises. How should responsibility and authority for quality be positioned within the governance structure? If it is treated purely as a technical or administrative task, it may be buried within a support function with limited influence. If, on the other hand, it is treated as a management responsibility, it must be connected directly to senior leadership and integrated into decision-making processes.

A common and practical solution is to establish a quality committee. The purpose of this is to ensure coordination, oversight, and follow-up on behalf of top management. The quality committee typically includes senior managers and key representatives from the core functions. Its existence sends a clear signal that quality is taken seriously and has a legitimate place in the governance structure.

Closely related to this is the role of quality manager. Making this role visible in the organogram is an important step in clarifying the responsibility. In many organisations, the quality manager reports directly to the managing director while working closely with the quality committee, as indicated by the broken line. This reflects the fact that quality management cuts across functional boundaries and requires authority that extends beyond a single department. At the same time, the quality manager does not own quality in an exclusive sense. Line managers, as well as managers of support departments, remain responsible for the quality of their own areas, while the quality manager acts as a coordinator, facilitator, and advisor.

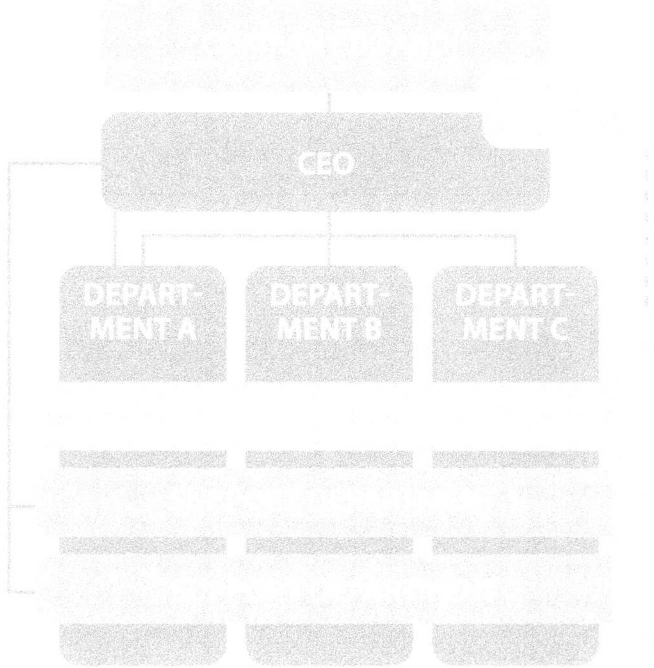

Figure 7.8 Example of a traditional organogram, where specific roles related to quality management are visible.

Middle managers play a particularly critical role in this regard. They translate the quality policy into daily practice, lead improvement efforts in their areas, and are responsible for responding to audit findings and deviations. If they do not take quality seriously, no formal QMS will function effectively. Therefore, their behaviour represents an important link between formal governance structures and everyday work. Moreover, a QMS that does not actively involve employees will most likely become disconnected from reality.

It is also important to recognise that governance structures evolve over time. During QMS implementation, more explicit roles and coordination mechanisms may be required. As quality management becomes embedded in daily practice and organisational culture, formal structures such as dedicated committees may become less central or be integrated into existing management routines.

Finally, governance structures should support the principle of professional autonomy, as discussed earlier. When responsibilities are well defined and authority is transparent, employees can better exercise professional judgement and take responsibility for the quality of their work.

7.7 Roles and competence in quality management

Clear governance structures make quality management visible, but structures alone are insufficient. Quality management ultimately depends on people performing specific roles, exercising judgement, and coordinating efforts across the organisation.

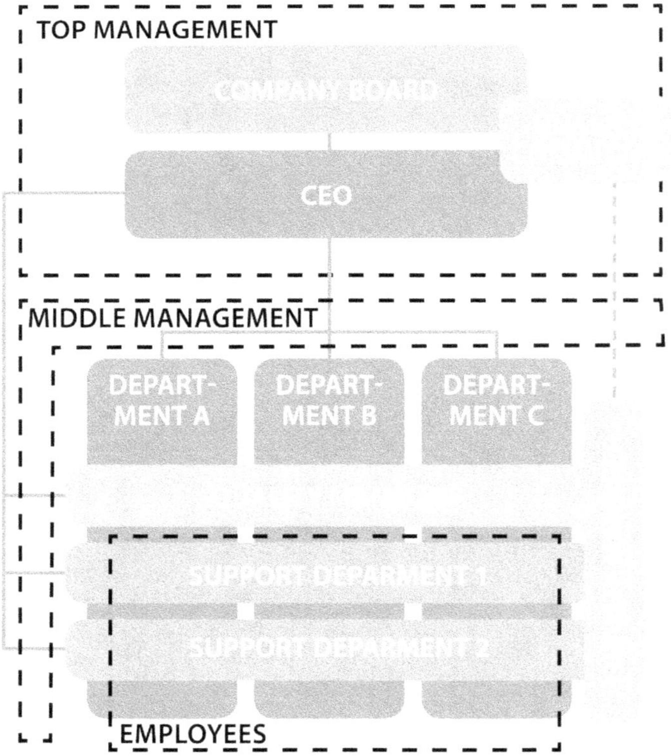

Figure 7.9 A presentation of organogram shows the roles that are important in the context of quality management.

A common misconception is that quality is owned by a single role, such as that of the quality manager. In practice, quality management is distributed across the organisation. Different roles contribute in different ways, and the effectiveness of a quality management system depends on how well these roles are understood, supported, and aligned with each other.

Figure 7.9 summarises how three groups within the organisation are crucial for the implementation and operation of a QMS, each in different ways. Top management carries overall responsibility. Middle management translates quality intentions into daily practice. Employees are closest to the work processes and are therefore best positioned to identify deviations, inefficiencies, and improvement opportunities.

The quality manager has a coordinating and facilitative role rather than a controlling role. In most organisations, the quality manager acts as an internal advisor to the management, supporting the development, maintenance, and improvement of the QMS. Typical responsibilities include coordinating documentation, facilitating audits, supporting improvement work, providing training, and acting as a link to external parties such as certification bodies. The quality manager rarely has direct authority over line operations; therefore, effectiveness depends largely on professional credibility, communication skills, and the ability to work across organisational boundaries.

In addition to these formal roles, many organisations rely on quality teams, improvement groups, and other cross-functional working groups. These are typically temporary and established to address specific issues or development tasks. Their strength lies in

Figure 7.10 The role of an independent consultant.

Figure 7.11 The house of competence of the quality manager.

their ability to bring together different perspectives and professional backgrounds. Such groups also support employee participation and contribute to a sense of shared ownership of quality management.

The final role worth mentioning is that of an independent consultant. An independent consultant does not have the same vested interests as internal actors and can, therefore, act as a constructive driving force in quality efforts. Consultants are often engaged when specific expertise is lacking, but also when the organisation seeks to build internal competence and give the development of the QMS additional momentum. Some consultants are highly visible and active during implementation, while others contribute at key milestones, supporting the quality manager and quality committee without broad organisational exposure.

Understanding these roles raises an important question: What kind of competence is required to manage quality effectively? Competence in quality management cannot be reduced to technical knowledge of standards or procedures. It also involves skills related to communication, coordination, leadership, and broader professional judgement.

To capture this more holistic view, we refer to a conceptual model known as the House of Competence of the Quality Manager (Ingason & Jonsdottir, 2017).[5] The model illustrates how competence is based on several interrelated components. Knowledge, including an understanding of quality principles, standards, process thinking, and improvement methods, lies at the foundation. In addition, skills such as organising work, facilitating collaboration, analysing data, and managing projects are required. At the highest level is competence, understood as the ability to apply knowledge and skills effectively in context—under pressure, in interaction with others, and in situations characterised by ambiguity and change.

Although the model was originally developed with the quality manager role in mind, its logic applies more broadly. Effective quality management requires competence at multiple organisational levels. Senior managers require strategic and leadership competencies related to quality. Middle managers require the ability to integrate quality into their operational decisions. Employees need sufficient understanding and professional autonomy in their daily work.

Roles and competencies are closely linked to the discussions in the earlier sections. A strong quality culture supports role clarity and competence development. Professional autonomy depends on knowledge, feedback, and authority, which are aligned with responsibility.

In the next section, we turn our attention to how these roles and responsibilities are made operational through communication, information flows, and documented practices, and how information systems support coordination and learning in quality management.

7.8 Information, documentation, and coordination

Quality management relies on structured communication. Policies, procedures, performance data, audit findings, and improvement actions must be accessible, understood, and followed. Without reliable information flows, coordination becomes fragile and learning remains accidental rather than systematic.

Information and records management are preconditions for a well-functioning quality management system. Research has shown that one of the areas where organisations struggle most during the implementation of a quality management system based on ISO 9001 is the handling of documented information (Gunnlaugsdottir, 2012).[6]

The requirements of ISO 9001 regarding documented information are sometimes perceived as being bureaucratic. In reality, they largely reflect well-established management practices for handling information in a structured and reliable way. Documents must be controlled, accessible, updated when necessary, and protected from unintended use. Records must be stored to ensure traceability and retrieval. These mechanisms make coordination and accountability possible.

Figure 7.12 The QMS consists of documents and records.

Documents and records

It is useful to distinguish between documents and records. Documents describe how work should be performed. These include policies, procedures, work instructions, templates, and process descriptions. They define expectations and provide guidance.

Records provide evidence of what has been done. Examples include signed contracts, completed checklists, audit reports, meeting minutes, performance measurements and corrective action reports. Records enable the review of performance, demonstration of compliance, and identification of improvement opportunities.

A QMS depends on both. Without documents, there is no shared reference for how work should be conducted. Without records, there is no evidence, traceability, or foundation for analysis and learning.

During implementation, many organisations underestimate the practical implications of document control. Consequently, several practical questions arise.

- Who is responsible for updating the procedures?
- How can we prevent the use of outdated versions?
- Where should the records be stored?
- How long should they be retained?
- Who has access?

In small organisations, these issues may initially appear manageable. However, as the number of documents grows, inconsistencies and duplications can occur. In larger organisations, complexity increases significantly owing to multiple departments, locations, and information systems.

The solution typically involves clarity, clear responsibilities, simple structures, and disciplined use. A small organisation may choose to maintain its QMS in a relatively simple format, perhaps as a limited set of digital documents stored on a shared drive or intranet. Updates can be made periodically, and the responsibilities for approval and revision can be clearly defined without creating an elaborate administrative structure. A larger organisation may require more formalised version control, defined responsibility for documents, and structured approval workflows. Responsibilities may be distributed across departments, and internal audits therefore require systematic traceability—meaning that documents, records, and decisions can be clearly linked, verified, and retrieved as evidence of how work has been carried out.

Need for specialised software?

It is commonly assumed that implementing a QMS requires purchasing specialised document management software. This does not have to be the case, and ISO 9001 does not prescribe how documented information must be managed; it merely requires that it be controlled.

The requirements of ISO 9001 can usually be fulfilled using existing information systems, if they are used consistently and responsibly. Shared drives, intranet platforms, and standard office software can be sufficient if:

- Responsibilities for document approval and revisions are defined.
- The access rights are clear.
- Obsolete versions are removed or archived.
- Records are stored in a structured and retrievable manner.

Investing in a new IT solution does not automatically resolve coordination problems. If employees do not use the system consistently or if roles are unclear, even the most advanced software will fail. Therefore, the focus should be on discipline and clarity, not technology. Simultaneously, as organisations become increasingly digital and interconnected, disciplined information management becomes even more important. Digital tools amplify both strengths and weaknesses in coordination; they do not replace the need for clarity of roles and responsible use of these tools.

Avoiding document inflation

One of the most frequent mistakes during implementation is document inflation, which is the uncontrolled growth of procedures, instructions, and forms. A formal documented procedure represents a commitment. Once something is written in the QMS, the organisation is expected to follow it. If practice diverges from documentation, nonconformities arise—not necessarily because performance is poor, but because documentation may be unrealistic, not sufficiently understood or known, or not consistently applied in practice.

The guiding principle should be necessity and value. Document what is needed to ensure consistency, compliance and learning. Avoid documenting commitments that do not contribute to quality or value creation.

Figure 7.1 illustrates the main reasons for creating quality documents in a simplified manner. In some cases, documentation is required by standards such as ISO 9001, where the organisation must provide evidence—typically in the form of records—that specific requirements have been met. In other cases, documented procedures are established to ensure operational consistency. Documentation may also be created for internal clarity, coordination, or strategic purposes.

However, caution is required. Promises made through documented procedures can have significant consequences. A work description was once published by a prestigious company with a certified QMS stating that each employee should receive a fleece sweater with the company logo every second year. Regardless of the original intention, ISO 9001 does not include such a requirement. However, once documented, it became an auditable commitment. If the sweater was not delivered, there was a nonconformity.

Therefore, companies should be careful about issuing unnecessary promises in their QMS that do not create value for customers or stakeholders.

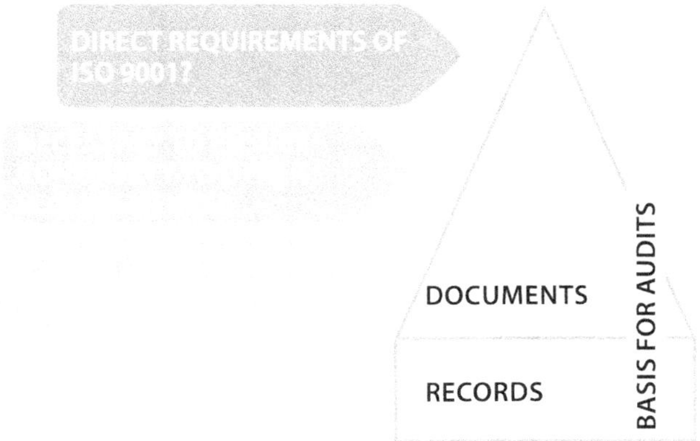

Figure 7.13 Assessing the need for a quality document.

Formal and informal information, communication, and transparency

It may be useful to distinguish between formal and informal information. Informal information may include drafts, internal notes, and working materials that have not yet been formally approved or issued. These do not always require full control under the QMS. Once information forms the basis of a decision, is issued to a client, or becomes part of a contractual or regulatory context, it becomes formal. At this point, traceability and retention requirements apply. Making this distinction explicit reduces unnecessary bureaucracy while maintaining control where it is needed.

Documented information supports quality management only if it is actively used. Policies must be communicated and understood by all parties. The objectives must be visible. Relevant performance information must be accessible to those who require it. Any deviations must be discussed. Audit findings must lead to appropriate actions.

In organisations with a strong quality culture, information flows are transparent. Information is used for learning rather than blame. This is where documentation connects directly to the broader systems perspective introduced earlier. Information flows are part of the organisational system that shapes behaviour and learning.

Gap analysis as a structured starting point

When an organisation intends to implement or revise a QMS according to ISO 9001 or another standard or reference, a structured gap analysis is a practical and disciplined starting point. Gap analysis involves comparing the current practice with the clauses of the reference. This can be done using a simple table.

If the reference is ISO 9001, the first column of the table can list the clauses of the standard. For each clause, the current practice is described, gaps are identified, and actions required to meet the requirements are defined. This approach provides an overview and

Figure 7.13 Gap analysis, comparing current practice with a reference.

highlights areas where the organisation already complies and areas where development is required.

Gap analysis can also reveal organisational and cultural issues, such as unclear responsibilities, weak information flows, or lack of performance data. In this way, gap analysis connects formal standard requirements with the broader organisational system described in this chapter.

7.9 Success conditions for organising quality in practice

Organising for quality involves culture, structure, roles, communication and leadership. A quality management system cannot function in isolation; it must be aligned with how people think, how decisions are made, how responsibility is understood, and how information flows.

Experience shows that certain conditions significantly increase the likelihood that quality management will become an integrated and sustainable part of organisational practice.

- Visible and consistent commitment from senior management is decisive. Quality must be reflected in priorities, resource allocation and leadership behaviour. Policies and objectives must be supported by follow-up, dialogue, and examples. Without active engagement from top management, even well-designed systems tend to lose their momentum. Consistent leadership, transparency, and disciplined follow-up also contribute to organisational legitimacy. In the long term, quality management supports trust among customers, employees, regulators and society at large.
- A realistic understanding of the organisational culture is essential. Before introducing formal structures or documentation requirements, it is necessary to understand the prevailing attitudes towards responsibility, transparency, and improvement. Tools such as quality culture discussions and structured surveys can support this understanding of quality culture. Cultural development must be supported through dialogue, participation and consistent leadership.
- The implementation should begin with a structured assessment. Gap analysis provides an overview of current practices in relation to recognised standards such as ISO 9001. This structured comparison highlights the strengths, weaknesses, and priorities for action. Importantly, such analyses often reveal cultural and organisational issues that must be addressed alongside technical gaps.
- Roles and responsibilities must be clearly defined and coherent. Quality is not owned by a single function. Senior managers, middle managers, quality professionals, and employees all contribute in different ways. Making these roles explicit through governance structures, organisational design, and competence development reduces ambiguity and strengthens accountability.
- Documented information must be handled with discipline and proportionality. The objective is to ensure traceability, coordination, and learning. Clear responsibility for documents, structured storage of records, and avoidance of unnecessary procedures are practical success factors.
- Employees need clear expectations, access to relevant information, and the ability to act when deviations occur. When responsibility, competence, and information are aligned, quality is built into daily work rather than being inspected afterwards. Governance structures and role definitions must support this principle.

- Quality management must be viewed as a dynamic process rather than a static structure. Organisations are open systems. They interact with suppliers, customers, regulators, and other stakeholders. Because requirements evolve and expectations change, learning must be continuous. A QMS provides a framework that supports adaptation and improvement over time.
- When culture, governance structures, and information flows are aligned, quality management functions as a stabilising capability in turbulent environments. It provides continuity without preventing adaptation and structure without suppressing initiative.

When these conditions are present, quality management becomes embedded in the organisation's way of working. It shapes communication, decision-making, and professional responsibilities. It contributes not only to compliance with standards but also to improved performance, lower costs, and stronger trust from customers and stakeholders.

In recent years, the discussion on quality management has also become more closely connected to the broader concept of social responsibility. Social responsibility refers to the expectation that organisations consider the wider consequences of their decisions and activities on society, the environment, and future generations. This perspective emphasises ethical behaviour, transparency, respect for stakeholders, and long-term thinking in organisational decision-making. Quality management can thus be understood not only as a means of meeting requirements and improving internal performance, but also as a way of strengthening trust and legitimacy in wider society. The international guideline ISO 26000 provides structured guidance on social responsibility. As noted in Chapter 4, ISO 26000 is not a certifiable management system standard, but it has nevertheless influenced how many organisations think about responsible and sustainable management practices.

Case: A certified system—and something still not working

A large printing company serving commercial and public clients had maintained a certified quality management system for several years. The system was formally structured, responsibilities were defined, and internal audits were conducted regularly. Documentation was updated, deviations were recorded, and management reviews were conducted as required. From a formal perspective, the company appeared to be disciplined and well-organised.

At the same time, the organisation was characterised by a strong professional culture. Employees took pride in their craftsmanship and valued flexibility in responding to customer needs. Production deadlines were often tight, and coordination depended heavily on direct communication and personal experience. When urgent adjustments were required, solutions were found quickly, sometimes without formal documentation.

Over time, small inconsistencies began to accumulate. Production planning did not always follow documented workflows. Adjustments to customer files were occasionally informally agreed upon. Certain inspection routines were completed in practice but were not recorded as prescribed. New employees noticed that written procedures did not always correspond fully with how experienced colleagues handled tasks.

Internal audits rarely revealed major nonconformities. Required documents were present. Templates were used. The corrective actions were formally closed. However, similar coordination problems continued to reappear across projects. The situation became more visible after a customer complaint exposed gaps between order confirmation, production adjustments, and final delivery schedules. The necessary procedures existed in the documented system. However, in practice, informal workarounds had gradually replaced parts of the prescribed workflow.

Management initiated discussions regarding whether the problem was insufficient control, insufficient documentation, or something else. Some proposed adding stricter monitoring. Others questioned whether documented procedures accurately reflect daily practice. Employees expressed mixed views; the system was not seen as harmful, but neither was it always perceived as helpful.

The company now faces a choice. It can strengthen formal control mechanisms and expand documentation to more closely mirror actual practice. Alternatively, it can reconsider how roles, responsibilities, communication patterns, and professional norms interact with the documented system.

Reflection points—Chapter 7

1. Culture versus compliance
   In a mid-sized construction company, formal ISO 9001 procedures are in place; however, experienced site managers often rely on informal coordination to meet deadlines. Should management tighten documentation and audit routines to ensure compliance, or revise procedures to better reflect real practice? How would each choice affect professional responsibility, learning, and long-term credibility? Discuss the trade-off between formal conformity and practical capability.
2. Professional autonomy or supervision
   Consider a hospital department where the work is knowledge-intensive and time-sensitive. Is quality better secured through close managerial oversight or professional autonomy supported by competence and transparent feedback? What risks arise if autonomy is high but competence is uneven, or if supervision is strict but trust is low? Reflect on the balance between control and professionalism.
3. Governance structure and real power
   In a public-sector organisation, quality is formally assigned to a support unit with limited authority. Should it be repositioned closer to top management or integrated more strongly into line responsibility? What signals would each structural choice send regarding priorities and legitimacy? Discuss the tension between the formal placement for responsibility for quality in the organisational structure (and the signals this sends) and its actual influence in practice.
4. Document simplicity or digital integration
   A growing engineering consultancy is considering investing in an advanced document management system to support its quality management system (QMS). Would improved digital integration strengthen coordination and traceability, or would the disciplined use of existing tools be sufficient? How does organisational maturity influence this decision? Discuss the trade-off between technological sophistication and behavioural discipline.
5. Quality culture and sustainability.
   In a logistics company under pressure to reduce costs, quality initiatives focus primarily on defect reduction and improving efficiency. How might a broader quality culture

incorporate environmental impact and social responsibility without diluting the operational focus? Reflect on the trade-off between short-term performance indicators and long-term legitimacy.

6. Agility and stability
   A software development team operates using agile methods and self-organising Scrum practices, while corporate management requires strict documentation for compliance reasons. How can an organisation reconcile adaptability and speed with stability and accountability? What are the minimal structures necessary to support learning without undermining agility? Discuss how quality management can stabilise change rather than constrain it.

Notes

1 Senge, P.M. (1990). *The Fifth Discipline: The Art and Practice of the Learning Organization.* New York: Doubleday.
2 Johnson, G., Scholes, K., & Whittington, R. (2008). *Exploring Corporate Strategy*, 8th ed. England: Prentice Hall.
3 Cameron, K., & Sine, W. (1999). A framework for organizational quality culture. *Quality Management Journal*, 6, 4.
4 Goetsch, D.L., & Davis, S. (2012). *Quality Management for Organizational Excellence: Introduction to Total Quality*, 7th ed. USA: Prentice Hall.
5 Ingason, H.Þ., & Jonsdottir, E.R. (2017). The house of competence of the quality manager. *Cogent Business & Management*, 4(1), 1345050.
6 Gunnlaugsdottir, J. (2012). Information and records management: A precondition for a well functioning quality management system. *Records Management Journal*, 22(3), 170–185.

8 Control of processes

8.1 From implementation to operation

For many organisations, the implementation of a quality management system is a project that they can understand. It has a defined scope, timeline, and often a clear milestone, sometimes even an external recognition or certification at the end. This can create energy and momentum. People see a goal, and the organisation works towards it together. Operating the system is different. There is no final line; it is something that continues as an integrated part of everyday routines. Maintaining momentum, keeping people motivated, and ensuring that the system remains alive in everyday work can be more demanding than the implementation itself. If the idea behind the QMS was never fully understood or "sold" within the organisation, there is always a risk of falling back into old routines once the initial project energy disappears.

Figure 8.1 The operation of a QMS, long walk ahead.

DOI: 10.4324/9781003671749-8

Measurements are needed to assess whether processes are active and effective. Processes must also be controlled to ensure that they deliver outputs in accordance with requirements. Internal audits must be prepared and executed, and management must regularly review the system to ensure that it remains suitable and effective.

The word *control* can sometimes sound rigid or authoritarian. However, in the context of quality management, control is not primarily about authority but rather about awareness and responsibility. In many cases, it is simply about professional autonomy. If structural steel beams are produced, the control may involve precise measurements of dimensions and load-bearing capacity. In this case, the control is technical and strict by necessity. However, control can also be much less technical. If you are working in a home for elderly people, providing personal care, food, and daily support, control may simply mean being aware of whether the residents receive what they need and whether the service is respectful, timely, and consistent. This means asking the simple question, are we doing what we promise to do? Are we meeting the needs of those we serve?

Control, in its simplest form, is about ensuring that what we deliver corresponds to the agreed requirements and expectations. Management involves clear observation and follow-up. Without some form of measurement or reflection, it is merely guesswork.

The purpose of this chapter is to explain how organisations create this awareness in a structured way, how they define what matters, measure it sensibly, understand variation, and use that understanding to maintain stable and reliable processes.

8.2 Quality control as a pillar of quality management

Earlier in this book, reference was made to Juran's quality trilogy, which describes quality management as consisting of three fundamental activities: quality planning, quality control, and quality improvement. Quality planning involves defining processes and setting objectives. Quality improvement involves making systematic changes that lead to better performance. Between these two stands quality control.

Quality control is concerned with maintaining performance within the defined limits. It is a discipline that ensures that processes operate as intended and that the results meet the established requirements. It involves monitoring whether a process operates within certain criteria or control ranges. The criteria should be consistent with the quality objectives and reflect the needs and requirements of customers. They must be measurable, fair, and understandable. The measured results are compared with these criteria to determine whether deviations are significant.

To illustrate this in a simple way, imagine you run a company producing dried fish, a popular Icelandic snack, which is shipped to supermarkets across the country. The packaging states that it contains 200 g of dried cod. This is an expensive product, and the declared weight is therefore important. As a responsible producer, you do not want to disappoint your customers. The criterion in this case is clear, each package should contain 200 g. If the weight is below this value, the customer receives less than what was promised. If it is consistently above, the company is giving away product. Quality control in this case means monitoring whether the packaging process keeps the weight aligned with what is promised on the label.

There may be a difference between the measured results and the defined requirement for two reasons. This may have a specific explanation, such as a malfunction or error, or it may be due to the inherent variability in the process. Distinguishing between these two

Figure 8.2 Selling 200 g dried cod pieces.

situations is at the heart of quality control. It is about understanding the behaviour of processes over time and ensuring that they remain stable and predictable.

8.3 Measurement as a precondition for control

Measurement is a prerequisite for quality control. In the absence of reliable data, there is no sound basis for assessing whether the processes are performing as intended. Therefore, a fundamental question in control is what should be observed. There may be one or more parts of the operation that are critical. These must be identified clearly, and appropriate indicators must be defined that reflect what truly matters.

The process must be monitored, and the results must be compared with defined references or specifications. When a value falls outside the specifications, it indicates that something is wrong, and action should be taken. Indicators should be designed carefully, focusing on what is important and avoiding the trap of measuring only what is easy to measure. A key question is how the results will be used. If they do not influence decisions or actions, or contribute to traceability, they may not be necessary.

Both internal and external customers should be considered when choosing what to measure. When employee-related aspects are observed, it is necessary to explain the purpose and involve employees in the preparation and implementation. Data must also be collected at the right points in the process, for example, when changes are made to the product during production.

Figure 8.3 Capelin processing and measurement of critical parameters.

The system for monitoring performance should be easy to use and understand. Indicators must be reassessed regularly and may be discontinued if they are no longer relevant. Collecting information is not enough; it must lead to action when necessary.

Measurements and observations are related to the elements of the operation that are to be kept within defined limits. The factors that influence the characteristics of the product or service must therefore be identified. These characteristics, in turn, affect how customers perceive the offering—including their satisfaction and loyalty. One way to identify such factors is to ask customers how they rate the product or service they receive.

Imagine you are responsible for processing capelin, a valuable pelagic fish that is often frozen for export to the Japanese market. Buyers expect large, well-formed fish with good texture and appearance. Size, maturity and quality directly influence the price. Capelin, as with other fresh fish, needs to be processed quickly to ensure the best experience for the consumer. From the moment they are caught, time becomes critical. One important parameter is the time from catch to freezing. If too much time passes, the product will deteriorate. Processing must also be timed correctly in relation to rigor mortis, as this affects the texture and shape. Therefore, monitoring time at defined stages of the process is essential. Another critical variable is the storage temperature. Because capelin contains fat, an incorrect temperature can quickly reduce its quality. Continuous monitoring is necessary to ensure that products remain within defined limits. In this case, the control is about observing the process at the right moments when the product is changing. Time and temperature are followed for several reasons; for example, they can be used during production and determine whether the final product meets market requirements.

8.4 Designing measurements and defining variables

When defining what to observe, it is useful to recall the discussion in Chapter 1 on quality characteristics and quality features. Quality can be viewed both in terms of the characteristics designed into a product or service (quality of design) and in terms of how consistently those characteristics are delivered in practice (quality of conformance). These two perspectives provide important context for measurement. Both are relevant for quality control and must be reflected in the defined indicators.

It is also useful to distinguish between the two types of variables. Lagging variables describe what has already happened. They are useful for evaluating performance over a defined period and for comparing results over time. Financial results, such as operating profit or sales income, are typical examples. Such information can be important for many stakeholders, such as the Company Board and the executive team, who are expected to develop a new plan to increase sales in the next quarter if the report from the last quarter is disappointing. However, lagging variables cannot be used in the continuous control of a process because they do not allow timely intervention.

In contrast, leading variables can be observed during the process itself. Their results can be used to make immediate adjustments, thereby influencing the outcome. An example would be an aluminium plant casting ingots with a defined manganese (Mn) content. These are often specialised products designed for enhanced strength and corrosion resistance. In this case, the chemical composition of the alloy must be measured prior to casting. If the composition does not comply with the client's requirements, manganese can be added to the melt before casting. Therefore, the variable makes it possible to correct the process before the final product is completed.

Figure 8.4 The last child picked up from kindergarten, 25 minutes too late.

In many cases, the quality characteristics of a product or service cannot be measured directly in real time, and sometimes not even after the process is completed without significant cost or delay. In such situations, control is achieved by monitoring process-related indicators that influence the outcome. For example, in food distribution, it may not be possible to assess product condition continuously, but maintaining the required temperature during storage and transport serves as a proxy for preserving product quality. Such indicators allow organisations to control what matters by focusing on variables that are observable and actionable during the process.

Another example from a service provider could be a kindergarten, where the regular working day finishes at 17:30, at which time all the children need to be picked up. A leading indicator could be whether parents inform staff in advance if they are delayed. This allows the organisation to plan staffing and manage the situation before it leads to overload or additional costs due to overtime work.

Examples of subjects for more technical process control could be the temperature of geothermal water in a district heating system, throughput of a fish filleting machine, or dust emissions from the filter of a silicon plant. For a service company, relevant examples could be the response time when handling orders in a fast-food restaurant, the speed of responding to customer inquiries in a service centre, or the number of people waiting in a queue in a supermarket.

The unit of measurement is important when defining indicators. Everyone should understand what is being measured and how the results are interpreted. This can be simple when measuring whether a product complies with the requirements. In this case, an "instance" may refer to a nonconforming unit. The measurement may then be expressed in the form:

Number of instances / highest possible number

For example, the frequency of defects in production:

Defect units / manufactured units

Figure 8.5 Measuring the durability of an armchair.

In industrial production, it is often crucial to keep production running most of the time, meaning that equipment remains in operation and downtime is minimised. To monitor this, one could define an indicator such as:

Operating time / total time

Product characteristics can be difficult to measure; however, the following provides a practical example. At IKEA, customers can sometimes witness the durability testing of manufactured goods—for example, how many times one can sit in an armchair before it gives way. Regardless of the method chosen, the measuring device and procedure must provide results in the agreed form and unit that has been defined.

8.5 Variation and the logic of control

In Chapter 2, Shewhart's fundamental insight was introduced: all processes exhibit variation, and not all variations should be treated in the same way. He distinguished between common and special causes. This distinction is central to the quality control process.

Variation simply means that the results are not identical every time a process is performed. Even under stable operating conditions, small differences occur. Machines warm up and cool down, materials vary, environmental conditions change, and human performance is never perfectly uniform. These unavoidable fluctuations are referred to as common cause variation. They are inherent to the process itself and reflect its normal behaviour.

Common cause variation is predictable within certain limits. If the process is stable, the variation follows a pattern that can be observed over time. Reducing common-cause variation usually requires changes in process design, equipment, or operating conditions. It cannot be changed by reacting to the individual outcomes.

Special cause variations arise from specific and identifiable events. Examples include equipment malfunctions, incorrect settings, unusual raw materials, and human factors. Special causes disturb the normal pattern of the process and lead to results that differ significantly from what is expected. These deviations should be investigated and addressed. Special causes are sometimes referred to as assignable causes because they can be traced to a specific source.

This logic was illustrated earlier with the example of internet connection speed. Small fluctuations in download speed are rarely noticed if the service is stable. These represent common variation. When the connection suddenly becomes unusually slow or stops working, users immediately recognise that something is wrong. This represents a special cause.

Returning to the example of packaging dried fish, small fluctuations in weight are unavoidable and reflect common variation. However, if the weighing equipment becomes miscalibrated or a mechanical fault occurs, the weight may suddenly deviate significantly from its usual pattern. This represents a special cause that requires immediate attention.

Whether the variation is serious or not depends on the context and requirements. In some processes, small fluctuations are harmless and are to be expected. In other cases, the tolerance for deviation is extremely limited. The important point at this stage is whether we understand the nature of the variation. Acting on common variations as if they are special causes creates instability. Ignoring special causes allows problems to persist.

Figure 8.6 Internet outage: a special cause.

Reducing variation is not only a technical question but also an economic one. Improvements in process capability often require investments in equipment, process redesign, training or tighter controls. These efforts come at a cost, and the benefit depends on how much the customer values increased consistency. In some cases, customers are willing to pay for highly consistent products or services. In other cases, a certain level of variation is acceptable, and further reduction does not create additional value. Understanding this balance is therefore an important part of quality management.

Effective quality control depends on distinguishing between these two types of variation. Acting on common variation as if it were a special cause creates instability, while ignoring special causes allows real problems to persist. In the next section, we examine how statistical process control provides a structured and disciplined method for making this distinction in practice.

8.6 Statistical process control in practice

Statistical process control (SPC) is a structured method for measuring and analysing variation in a process. Its purpose is to distinguish between common and special causes of variation using statistically determined control limits.

At its core, SPC is based on a simple idea. If we understand the natural variation of a process, we can determine whether a new observation represents normal behaviour or signals that something unusual has occurred.

Control limits are calculated using probability theory. They represent the expected spread of the results when only common causes are present. A control chart does not identify the cause of a problem; it indicates that a special (assignable) cause may be present and that an investigation is required.

Figure 8.7 Monitoring the temperature and acidity in a public hot tub.

Ideally, only common causes should be present in the process. When all the plotted points lie within the upper and lower control limits and no unusual pattern appears, the process is said to be in a state of statistical control.

Figure 8.8 is an example of a simple control chart; the variable being controlled is temperature, and the graph shows how temperature changes with time. If the temperature

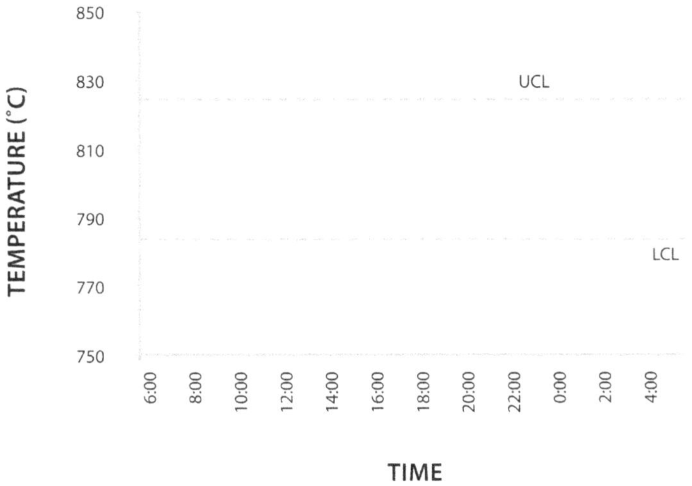

Figure 8.8 Control chart for temperature: UCL is upper control limit and LCL is lower control limit.

rises or falls below the control limits, this means that there is a specific problem in the operation, and immediate action must be taken.

For some readers, SPC may initially seem abstract or mathematical. In practice, this logic is not complicated. It provides a disciplined way of answering a practical managerial question: should we intervene or should we leave the process alone? In modern organisations, digital systems increasingly automate data collection and real-time monitoring, thereby enhancing the predictive power of statistical process control.

The ±3σ principle

In most control charts, the control limits are set at approximately three standard deviations ($\pm 3\sigma$) from the process mean. If the process variation follows a normal distribution, approximately 99.7% of observations will fall within $\pm 3\sigma$ when only common causes are present. A point outside these limits is statistically unlikely to be caused by normal variation alone.

This does not mean that the process is perfect. It only means that the probability of such a result occurring under normal conditions is very low.

Constructing an $\bar{X}$–R *control chart*

When a quality characteristic can be measured on a continuous scale, such as weight, temperature, length, or time, the most commonly used control chart is the $\bar{X}$–R chart. The idea is to monitor the process over time by taking small samples, called subgroups, at regular intervals. Each subgroup typically consists of a small number of observations under similar conditions. The purpose is to capture short-term variation within the subgroup and longer-term variation between subgroups.

For each subgroup, two values are calculated:

- The subgroup average $\left(\bar{X}\right)$.
- The subgroup range (R), defined as the difference between the highest and lowest measurement in the subgroup.

After collecting data for a sufficient number of subgroups, we can calculate:

$\bar{\bar{X}}$ = average of the subgroup averages
$\bar{R}$ = average of the subgroup ranges
These two values form the basis for the control limits.

Control limits for the $\bar{X}$ *chart*

The upper and lower control limits for the $\bar{X}$ chart are calculated as follows:

$$UCL = \bar{\bar{X}} + A_2\bar{R}$$

$$LCL = \bar{\bar{X}} - A_2\bar{R}$$

The constant A_2 depends on subgroup size. It is derived from statistical theory and reflects the relationship between the average range and standard deviation of the process.

For a subgroup size of $n = 5$, A_2 equals 0.577. This constant ensures that the control limits correspond approximately to ±3 standard deviations of the subgroup means.

Control limits for the R *chart*

The R chart monitors the variation within each subgroup. Its control limits are:

$$UCL = D_4\bar{R}$$

$$LCL = D_3\bar{R}$$

For subgroup size $n = 5$, $D_3 = 0$, and $D_4 = 2.114$.

The R chart ensures that the process variability is stable. It is good practice to first examine the R chart. If the variability is unstable, the interpretation of the $\bar{X}$ chart becomes unreliable.

Once the control limits are calculated, the subgroup averages and ranges are plotted over time. If all points lie within the control limits and no systematic patterns appear, the process is considered to be under statistical control. The observed variation is assumed to be a common variation.

If a point falls outside the control limits, this indicates that a special cause may be present. The correct response is investigation, not immediate adjustment of the process, but identification of the underlying cause. It is essential to avoid overreactions. Adjusting a stable process in response to common variations can increase instability.

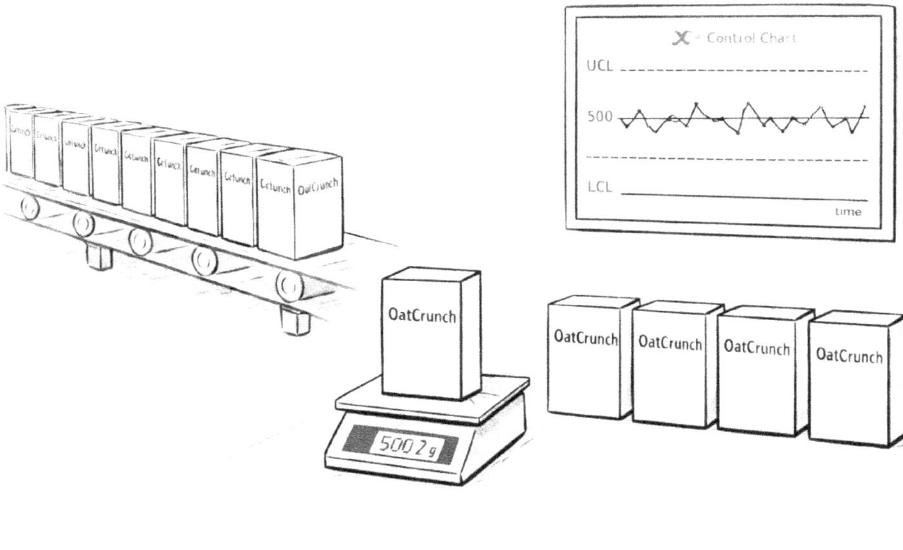

Figure 8.9 Developing a simple control chart for cereal production.

Applying an $\overline{X}$–R control chart in practice

The logic of an $\overline{X}$–R control chart is illustrated using a simple example. Consider a company that produces breakfast cereals. Each box is labelled as containing 500 g. Inspecting every box is costly and unnecessary. Therefore, the company decides to take a sample of five boxes every hour. These five measurements form one subgroup.

For each subgroup, the average weight $\overline{X}$ and the range (R) are calculated. Suppose that during one hour the measured weights are 498 g, 502 g, 501 g, 499 g, and 500 g.

The subgroup average is $\overline{X} = 500\text{g}$ while the subgroup range is $R = 502 - 498 = 4$ g.

This procedure is repeated each hour. After collecting, for example, 25 subgroups, the company calculates:

The overall average of subgroup averages: $\overline{\overline{X}} = 500.8g$

The average of subgroup ranges: $\overline{R} = 3.6g$

Using the previously introduced formulas for subgroup size $n = 5$, and the corresponding constants: $A_2 = 0.577$, $D_3 = 0$, $D_4 = 2.114$. The control limits for the $\overline{X}$ chart and R chart can now be calculated.

For the $\overline{X}$ chart:

$$UCL = 500.8 + (0.577 \times 3.6) = 502.88\text{g}$$

$$LCL = 500.8 - (0.577 \times 3.6) = 498.72\text{g}$$

For the R chart:

$$UCL = 2.114 \times 3.6 = 7.61$$

$$LCL = 0$$

These limits are then plotted together with the hourly subgroup averages and ranges.

If all subgroup averages fall between 498.72 g and 502.88 g and no systematic pattern appears, the process is statistically stable. The variation observed is then considered to be a common variation.

If a subgroup average exceeds these limits, this suggests that something unusual has occurred, such as a temporary malfunction in the filling mechanism, a change in cereal density, or an operator adjustment.

The control chart does not identify which of these events occurred. But this indicates that an investigation is required. Importantly, these control limits describe the natural behaviour of the process. They do not yet tell us whether the boxes meet customer or legal requirements. That question belongs to process capability, which we examine later in this chapter.

Zone testing and early detection of shifts

Traditional control charts are based on the $\pm 3\sigma$ principle; a point outside the control limits signals a likely special cause. This rule is powerful but is conservative. It reacts only to

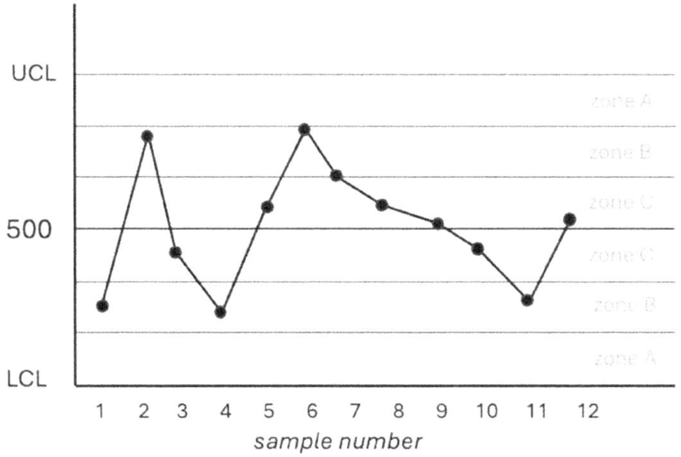

Figure 8.10 Zone testing, a complement to control limits.

relatively large deviations from the target. In practice, smaller but systematic shifts may occur before a point reaches the control limits. To detect such changes earlier, control charts are sometimes divided into different zones.

Starting from the centre line (the process mean), the chart is divided into three zones on each side:

Zone C: within ±1 standard deviation (σ) from the mean
Zone B: between $\pm 1\sigma$ and $\pm 2\sigma$
Zone A: between $\pm 2\sigma$ and $\pm 3\sigma$ (approaching the control limits)

This division does not change the control limits, but it adds structure to the interpretation. Zone testing introduces rules based on the patterns of consecutive points falling within these zones. Typical rules include:

- One point in Zone A may indicate that the process is moving towards instability.
- Two out of three consecutive points in Zone B or beyond on the same side suggest a possible shift.
- Four out of five consecutive points in Zone C or beyond on the same side may indicate a small but sustained shift.
- Eight consecutive points on the same side of the centre line indicate a likely change in the process level.

These rules are designed to detect non-random patterns—gradual drifts, shifts or trends—even when all points lie within the $\pm 3\sigma$ limits.

Zone testing should be considered a complement to traditional control limits, and the benefit of zone testing is earlier detection. A slow upward drift in the cereal filling process, for example, may not immediately exceed the control limits, but several consecutive points above the centre line may indicate that the average fill weight is increasing. If detected early, corrective actions can be implemented before waste accumulates.

Caution is required. The addition of too many rules increases the risk of false alarms. Every investigation consumes resources. Therefore, zone rules should be applied with discipline and interpreted in context.

Attribute control charts

Not all processes involve measurable continuous variables, such as weight or length. In many cases, quality is assessed by counting the number of defects.

Examples include the number of customer complaints, the number of defective units in a sample, or the number of nonconformities in an audit. In such cases, attribute control charts are used.

The *p*-chart monitors the proportion of defective units. The *c*-chart monitors the number of defects per inspection unit.

Although it is based on counting rather than measurement, the principle remains the same: distinguish between common and special causes.

Control limits describe the natural behaviour of the process and are calculated from observed data. Specification limits, by contrast, are defined externally by customer requirements. A process may be statistically stable and still fail to meet these requirements, or it may meet them while remaining unstable. This distinction becomes central when evaluating process capability, which is examined in the next section.

8.7 Process capability

A process in statistical control behaves in a predictable manner over time with only common causes of variation present. However, stability alone is insufficient. A process may be stable and still fail to meet the customer requirements. This leads to the concept of process capability. While control charts describe how a process behaves, process capability describes how well that behaviour fits within defined specification limits.

Control limits versus specification limits

Control limits are calculated from the data. They reflect the natural variation of the process. In contrast, specification limits are defined externally by customers, regulations, contracts, or design requirements.

Returning to the cereal example, suppose that the declared weight is 500 g, and legal or contractual requirements specify that the weight must lie between 490 g and 510 g. The specification limits are as follows:

Lower specification limit (LSL) = 490 g and upper specification limit (USL) = 510 g

Even if the control chart shows that the process is statistically stable, we must ask whether the natural variation of the process fits comfortably within these limits. This is the question that process capability answers.

The capability index C_p

The most basic capability index is

$$C_P = \frac{USL - LSL}{6\sigma}$$

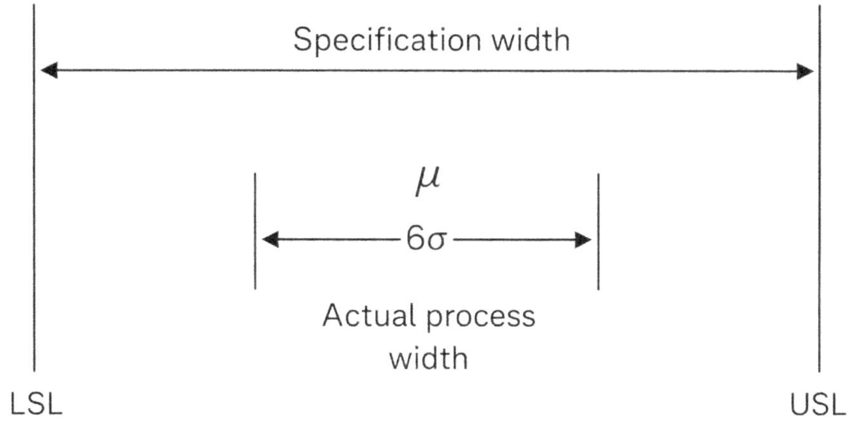

Figure 8.11 The basic logic of process capability (C_p).

σ is the process standard deviation. This formula compares the width of the specification interval with the natural spread of the process (approximately 6σ). If $C_p = 1$, the specification width is equal to the natural process spread. The process just fits within the limits.

If $C_P > 1$, the process variation is smaller than the allowed specification width.

If $C_P < 1$, the process variation is larger than the specification width. Even if the process is stable, some output will fall outside the specifications.

Importantly, C_P assumes that the process mean (m) is centred between the specification limits.

The capability index C_{pk}

In practice, processes are not always perfectly centred. A stable process may drift slightly towards one specification limit. To account for this, we use:

$$C_{pk} = \min\left(\frac{USL - \mu}{3\sigma}, \frac{\mu - LSL}{3\sigma}\right)$$

μ is the process mean and s is the process standard deviation. C_{pk} measures how close the process mean is to the nearest specification limit.

If the process is perfectly centred, $C_P = C_{pk}$. But if the process is off-centre, $C_{pk} =< C_P$.

Therefore, C_{pk} provides a more realistic measure of the actual capability.

Interpreting capability values

Higher C_{pk} values indicate a lower risk of producing nonconforming output. As a rule of thumb:

- $C_{pk} = 1.0$ means the process just meets specifications.
- $C_{pk} = 1.33$ is often considered acceptable in many industries.
- $C_{pk} = 1.67$ or higher may be required in more demanding contexts.

Returning to the cereal example

Suppose that the process for filling cereal boxes has $\sigma = 1.2$ g. This means that $6\sigma = 7.2$ g Because the specification width is 510 – 490 = 20 g, we obtain:

$$C_P = \frac{20}{7.2} \approx 2.78$$

This indicates a highly capable process, as the variation is small relative to the allowed tolerance.

However, if the mean has drifted upward to 508 g, C_{pk} will be significantly lower, even though C_P remains high. In such a case, the company may be giving away unnecessary products, increasing costs without increasing customer value. Therefore, process capability links statistical control to economic performance.

When variation tolerance is low

In some processes, the tolerance for variation is very small. Consider a pharmaceutical company that produces beta-blockers in tablet form for patients with heart disease. Dosage accuracy is critical. Regulatory requirements define extremely narrow specification limits. Even a small deviation may have serious consequences for the patient. In such contexts, very high capability indices are required. Stability is not enough; the process must be capable with a significant safety margin.

This example illustrates an important point: acceptable variation depends on the context. In some services, such as response times in a call centre, moderate variation may be acceptable. However, this is not the case in pharmaceutical production.

One important principle must be emphasised. Stability is a prerequisite for capability analysis. If special causes are present, the estimated standard deviation does not represent stable behaviour. In such cases, the capability indices are misleading. The process capability should only be calculated when the process is under statistical control.

If the capability is insufficient, there are two basic options. Reduce the process variation (reduce σ) or re-centre the process mean. Reducing variation often requires process redesign, equipment upgrades, training, or improved control methods. Re-centring may simply require a disciplined adjustment based on reliable data. In both cases, we return to the logic of improvement, which is addressed in the next chapter.

A brief note on "Six Sigma"

If a process is said to operate at a "Six Sigma" level, this typically means that the specification limits are positioned six standard deviations away from the process mean. Under the assumption of a normally distributed process and allowing for a long-term mean shift of 1.5σ, a convention introduced in Six Sigma practice, this corresponds to approximately 3.4 defects per million units. In other words, the probability of producing a non-conforming unit is extremely small.

Achieving this level of performance can be expensive. It requires highly stable processes, disciplined measurement systems, and continuous improvement. Such performance may be appropriate in contexts where failure has serious consequences, for example, in pharmaceutical production or safety-critical components. However, not all processes require this level of precision. The economically optimal capability level depends on the context, risk involved, and cost of defects.

So far, we have focused on controlling variation within individual processes. But quality management also requires control of the system itself. It is not enough that processes are stable; we must also ask whether procedures are followed and whether the system remains effective over time. Internal audits and management review provide this broader level of control.

8.8 From review to internal audit

In Chapter 6, we described how an engineering consultancy manages its projects. Each project is formally defined, planned, and executed. Before delivery, the output is reviewed internally. Only after this review is the project delivered to the client and formally closed.

This internal review is not an audit in the formal sense. It is simply good practice and reflects professional responsibility. Before handing over the work to the client, the consultancy checks whether the outcome meets expectations. The review is carried out by a specialist who was not directly involved in the project, to ensure an independent assessment. Such review mechanisms demonstrate that the organisation takes responsibility for examining its own performance.

The idea of an internal audit is simply a structured and systematic extension of a very natural principle: "We periodically review whether we are doing what we have agreed to do."

Review as a control mechanism

Reviews in engineering consultancy can take many forms in practise.

- A project review checks whether the output meets the requirements before delivery.
- A peer review examines the technical calculations.
- The supervisor may review the documentation before approval.

All of these are forms of internal control. They aim to detect deviations before they reach customers.

This idea can be easily expanded. In every organisation, general discussions and exchanges of experience take place regarding how tasks are performed. Constructive criticism and dialogue are natural aspects of professional life. Questions such as "Why do you do it like this?" can trigger reflection and improvement.

Consider an organisation that suddenly needs to plan and execute many internal projects to adjust its operations to new requirements. First, there may be no formal approach. Project managers prepare plans differently based on their individual experiences. Over time, however, they realise that they are stronger as a group. By sharing approaches and discussing good practices, they gradually established more consistent and effective ways of working. This too is a form of review—informal, experience-based, and rooted in professional dialogue.

Figure 8.12 Internal audits and the quality pyramid.

When an organisation implements a formal QMS, this idea is further expanded. The organisation not only reviews individual outputs but also examines whether its defined processes and procedures are being followed consistently.

Internal audits in a formal quality management system

In organisations that follow standards such as ISO 9001, internal audits are a required and structured form of review. An internal audit is an independent assessment of whether:

- The organisation follows its documented procedures.
- The management system complies with the requirements of the chosen standard.
- The system functions effectively.

Figure 8.12 illustrates the principal idea behind internal audits in the context of the quality pyramid, which has served throughout this book as a symbol of the quality management system. In the upper part of the pyramid are the formal quality documents, quality manual, policy, and defined procedures. Records form the foundation of the system. These records provide evidence that the requirements of the QMS have been fulfilled. They may include meeting minutes, checklists, reports, documented communications, and other forms of traceable documentation.

The magnifying glass in Figure 8.12 represents an internal audit. It examines both the upper part and the foundation of the pyramid, the documented intentions, and the evidence that these intentions have been implemented in practice.

One important condition should be noted. An auditor must not audit his or her own work. The purpose is to obtain an objective picture of how the system operates. Therefore, internal audits address two fundamental questions. Are we doing what we say we are doing? Does our management system function as intended?

The nature of internal audits

In organisations with a healthy quality culture, audits are constructive. Internal auditors collect evidence and document facts. They identify nonconformities where requirements are not met and may also highlight opportunities for improvement. While audits serve an important control function by ensuring that the system operates as intended, they also contribute to organisational learning, the sharing of good practices and the identification of areas for further improvement.

Auditors should be aware of their role in identifying systematic weaknesses in quality management systems. Considerable resources are typically invested in designing and implementing the system. Therefore, it is in everyone's interest to identify weaknesses and improve them over time. Auditors must not see themselves as law enforcers trying to "catch" individuals who violate the law. Deviations rarely arise because employees deliberately ignore the procedures. More often, they reflect unclear or outdated instructions, insufficient explanations, or processes that are unnecessarily complex. Such issues are typically addressed through training, improved communication, and simplification of the system itself.

A committed group of internal auditors who meet regularly can become one of the most important collaborative forums in the operation of QMS. Through structured dialogue across departments, they contribute to transparency, shared understanding, and organisational learning. By focusing on root causes rather than symptoms, internal audits help prevent recurring problems and support preventive actions.

Internal audits are central to maintaining the integrity of the quality management system. In fact, it can reasonably be argued that internal audits are the "engine" of an active QMS. They prevent the system from becoming a static set of documents and help ensure that it remains active and relevant in everyday operations.

Finally, by ensuring transparency, traceability, and factual assessment, internal audits support the credibility and legitimacy of the organisation. They demonstrate to stakeholders that the management system is actively monitored and maintained.

The internal audit process

Guidance on how to conduct management system audits is provided in ISO 19011 (ISO, 2025),[1] the international standard for auditing management systems. The standard outlines the principles of auditing—including independence, evidence-based assessment and professional judgement—and describes how audit programmes are planned, executed and followed up. This is aligned with the plan–do–check–act (PDCA) cycle and supports a structured, risk-based approach to auditing. ISO 19011 provides useful guidance for designing and conducting internal audits in a consistent and professional manner.

An audit plan is usually prepared for a defined period, for instance, one year. In many organisations, the quality manager prepares the plan. The plan should specify which areas or processes will be audited, which requirements are being examined, when an audit will take place, and who will conduct the audit. Occasionally, audits focus on specific work procedures. In other cases, they focus on a particular requirement in the reference standard.

Each internal audit is then prepared and conducted by the assigned internal auditor, who is also responsible for reporting the results of the audit and for any follow-up and pushing for subsequent improvements.

Preparation and desk review

Preparation begins by reviewing the audit plan and conducting a desk review based on the documentation and records.

Circumstances change continuously, and organisational needs evolve over time. Therefore, the quality management system must be adapted over time. An important purpose of the audit is to assess whether the system remains relevant and is up to date. The desk review involves examining relevant procedures and associated documentation. The auditor checks whether

- The documented procedures meet the requirements of the reference standard.
- The documentation reflects actual practise.
- The system has been updated where necessary.

The term desk review reflects the fact that this part of the work can be done at the auditor's desk without yet engaging directly with the auditee. During this stage, auditors typically prepare a checklist. The checklist may be based on a general template provided by the quality manager. It should include important dates and identification details, relevant reference requirements, and areas to be examined. There should be space for writing notes and comments.

Good preparation is essential. The auditor should be clear about the sequence of topics and the time allocated to each. Audits should be conducted efficiently and must not become unnecessary disturbances to daily operations. Once preparation is complete, the auditee is informed about the audit, its purpose and expected duration, and an agreement is reached on timing.

Conducting the audit

During the audit, the focus must remain on whether actual practice complies with documented procedures and defined requirements. It is important to keep to the planned scope and schedule. The temptation to engage in general discussions on unrelated aspects should be resisted. Audits are structured activities with defined objectives.

If a deviation from the documented procedure is identified, it is recorded as a nonconformity. The auditor explains the observation and seeks agreement with the auditee regarding the factual description, classification of the nonconformity, and wording used in the report.

Only facts, not assumptions or personal interpretations, should be recorded. The focus is on the process rather than on individuals. Names of employees should not be included.

Reporting and follow-up

After the audit, the auditor reviews all collected information and summarises the findings. Both positive observations and nonconformities should be highlighted. Careful wording is important, as many people may read the report.

An audit report is prepared and presented to the responsible manager. The report provides a factual overview of the findings and serves as the basis for the corrective actions. The responsible manager is given a defined period to implement the necessary corrective actions. Follow-up is then conducted to verify that corrective actions have been implemented and are effective.

8.9 Management review

In most organisations, there are ongoing internal discussions about whether the way things are done is the best possible way. Are the procedures efficient, economical, and safe? Do they meet the customer requirements? Can they be simplified or improved?

At the highest level, the Company Board examines annual reports and holds the CEO accountable. Questions are raised about performance compared to previous years and

Figure 8.13 Management review.

in relation to competitors. Why have the results improved or declined? What explains changes in customer satisfaction, cost levels, or profitability? Such questions are not only financial; they often point to the way operations are organised and managed.

Similar discussions take place at different levels of the organisation. The management team reviews the operational results and considers adjustments. Departmental meetings address recurring problems and coordination challenges within the department. Process owners discuss how responsibilities are defined and whether the procedures function as intended.

In organisations with a strong quality culture, such reflection is natural and a part of responsible management. This reflects an awareness that performance must be monitored and that established ways of working should not be taken for granted.

The concept of *a management review* is nothing more than a structured and formalised version of a very natural practice. Namely, periodically stepping back to assess whether the organisation is performing as intended and whether its management system remains appropriate.

In organisations that follow ISO 9001, management reviews are a formal requirement. Top management must periodically review the QMS to ensure that it remains suitable, adequate, and effective. However, beyond compliance with standards, management review is fundamentally a leadership responsibility.

If internal audits ask, "Are we doing what we said we would do?" management review asks, "Are we doing the right things, and are they still relevant?" It provides a structured opportunity for senior leadership to consider internal audit results, process performance and capability, customer feedback, nonconformities and corrective actions, risk and opportunity assessments, progress towards quality objectives, and resource adequacy. The purpose is to assess whether the QMS supports the organisation's strategic direction and whether adjustments are necessary to ensure its effectiveness.

Figure 8.14 Management system as an ongoing practise.

A quality management system must evolve with the organisation. Markets change, technologies develop, regulations are updated, and strategic priorities shift. A management review is a mechanism that helps ensure that the QMS adapts to this changing context rather than gradually losing its relevance.

Earlier in this chapter, we described control at the operational level using statistical process control. We then examined internal audits as a system-level control. Management review represents control at the strategic level of the organisation. This ensures alignment between policy and objectives, operational performance, customer expectations, and overall direction. A system may be stable and compliant but no longer aligned with strategic priorities. Management reviews prevent such misalignments from developing unnoticed. Its effectiveness depends on genuine leadership engagement. When top management actively participates by questioning data, analysing trends, and challenging assumptions, management review becomes a powerful instrument.

Management reviews may also include consideration of broader societal expectations, sustainability objectives, and long-term stakeholder value. In this way, the quality management system can support not only operational performance but also responsible and future-oriented conduct.

Therefore, quality management operates at several interconnected levels, including statistical control of processes, internal audits of the system, and management review of overall performance and direction. Together, these mechanisms create a structure that supports stability, accountability, and alignment.

There is no finish line. The activities described in this chapter are recurring responsibilities. They are part of managing quality in an ongoing and disciplined way.

Case: Controlling stability in a public utilities company

A publicly owned utility company responsible for water distribution and wastewater treatment operates in a highly regulated environment. Service interruptions are unacceptable, environmental requirements are strict, and public trust depends on consistent performance. The company had implemented a certified quality management system several years earlier and had gradually strengthened its measurement and follow-up routines.

The operational processes were continuously monitored. The water quality indicators were measured daily. The treatment parameters were recorded automatically. Maintenance activities were scheduled at predefined intervals. Defined specification limits were established in accordance with regulatory requirements, and deviations beyond these limits triggered immediate corrective actions. The organisation appeared to be technically well controlled.

Over time, the company introduced statistical monitoring of selected parameters to better understand process variation. Control charts were gradually implemented for critical variables such as turbidity and chemical dosage. This allowed supervisors to distinguish between values that exceeded the regulatory limits and those that fluctuated within the statistically expected ranges.

In addition to operational measurements, the company tracked broader performance indicators, such as the response time to incidents, number of customer complaints, maintenance backlog, and compliance with regulatory reporting deadlines. These indicators were reviewed during regular management meetings. Internal audits were conducted according to an annual plan, and the findings were documented and followed up systematically.

Despite this structured approach, the management faced recurring challenges. Certain indicators fluctuated significantly from month to month without an obvious explanation. Some deviations were traced to identifiable causes, such as equipment malfunction or extreme weather conditions, which are clear examples of specific causes. Others remained within the control limits but still created concerns among managers.

When incidents occurred, the immediate response was usually effective. Corrective actions were implemented, and the documentation was updated. However, discussions increasingly centred on a deeper question: Were managers reacting to special causes or to normal variation? Did every fluctuation require an intervention? Were some corrective actions taken to increase instability rather than reduce it?

Internal audits confirmed that the data collection routines were largely followed. However, they also revealed differences in their interpretations. In some departments, minor variations within the control limits triggered adjustments. In others, similar patterns were considered to be inherent process behaviours. The system was consistent in documentation, but less so in statistical reasoning.

During management review meetings, discussions moved beyond compliance. Leaders began to question whether the selected indicators truly reflected long-term stability and societal expectations. Were their measuring capabilities relative to the regulatory requirements? Were they over-controlling certain processes while neglecting others? Were resources used to "correct" variations that were statistically normal?

The company's experience demonstrated that control is not achieved simply by collecting data or conducting regular audits. Stability depends on understanding variation, designing meaningful measurements, and exercising disciplined judgement in interpreting signals over time.

A central dilemma remained: How should management decide when to intervene and when to allow a stable process to operate without adjustment?

Reflection points—Chapter 8

1. Intervention or patience
   In a manufacturing company producing packaged food, control charts show repeated fluctuations within control limits but no points outside them; however, supervisors feel uneasy and propose adjustments to tighten performance. Should management intervene to reduce visible variation or allow the statistically stable process to operate undisturbed? Reflect on the trade-off between responsiveness and over-control and analyse how unnecessary adjustments may increase instability rather than improve capability.
2. Compliance or capability
   A provider of banking services company meets regulatory specification limits but operates with a C_p close to 1.0, leaving little margin for unexpected shifts. Should leadership invest in reducing variation further despite full compliance or allocate resources elsewhere? Analyse the trade-off between minimum compliance and long-term robustness and consider how risk tolerance differs across sectors.
3. Leading or lagging signals
   In a private healthcare clinic, management primarily reviews annual financial results and patient satisfaction surveys, whereas process-level indicators, such as waiting time variability, are rarely analysed. Should attention shift towards leading operational variables even if financial outcomes appear satisfactory? Reflect on the trade-off between short-term results and early warning signals in service-based organisations.
4. Standardisation or professional judgement
   In an engineering consultancy, internal audits revealed differences in how project managers interpreted procedural requirements. Should the organisation enforce stricter standardisation or preserve flexibility for professional discretion? Discuss the tension between consistency and adaptability in such an organisation.
5. Control signals or ignored feedback
   In universities, students frequently complain about late course syllabuses and delayed grading. Course evaluations are collected after each course; however, many students believe that the feedback is rarely used to improve teaching or administration. Should university management treat these as isolated issues or introduce a systematic process for analysing patterns and corrective action? What are the risks? Reflect on the trade-off between departmental autonomy and building a reliable system of control and improvement that strengthens student trust.
6. Stability or strategic adaptation

During a management review in a technology company, senior executives recognise that major market changes are approaching due to new regulations and emerging digital competitors. The existing quality management system has helped create stable and predictable operations, but it was designed for the current business model rather than the one that may soon emerge. Should leadership rely on the existing QMS to guide the transition, adapt it to increase flexibility, or consider a more fundamental redesign of the system? Reflect on the trade-off between preserving stability and enabling strategic transformation.

Note

1 International Organization for Standardization. (2025). ISO 19011:2025 guidelines for auditing management systems. ISO.

9 The continuous quest for becoming better and better

9.1 Continuous improvement as the third pillar: Organised as projects

In the previous chapter we discussed control. We observed how measurement makes control possible and how stability in processes is a prerequisite for delivering products and services according to defined requirements. However, stability alone is insufficient. A process can be stable and inefficient. It can consistently deliver results that may be acceptable, but not competitive. If organisations accept this situation, they will gradually lose their ground.

Juran's quality trilogy reminds us that quality management rests on three pillars: quality planning, quality control, and quality improvement. Planning establishes goals and defines the processes required to achieve them. The control ensures that the processes operate within the defined limits. Improvement is what moves the organisation forward; it is the deliberate effort to raise performance to a new level. Continuous improvement is a fundamental management responsibility of the organisation. Markets change, technologies evolve, stakeholder expectations increase, and new risks emerge. What is good enough today may not be good enough in the future.

Continual improvement is embedded in the international management standards discussed earlier in this book. ISO 9001, like other ISO management standards, is built around the plan–do–check–act cycle, where "act" explicitly requires organisations to continually improve their processes and management systems. These standards describe a dynamic management system that must evolve. The system is expected to drive improvements.

Improvement is closely related to learning. Organisations learn from deviations, incidents, customer feedback, and changes in their environment. When problems occur, they should not only be corrected; the causes should be identified and eliminated to prevent the reappearance of the same problems. Simultaneously, improvement is not limited to reacting to failures. Preventive thinking is equally important in this regard. By analysing trends, assessing risks, and monitoring performance, organisations can anticipate weaknesses before they develop into serious problems.

In this book, quality management is consistently viewed from a project management perspective. Improvement work fits naturally into this perspective. The most effective way to deal with improvements is to organise them as a project. Improvement initiatives have a clear objective, defined scope, responsible persons, measurable goals, and a time frame. They are planned, executed, and followed up. When the desired result is achieved, the new way of working must be anchored in the management system so that the improvement becomes permanent.

DOI: 10.4324/9781003671749-9

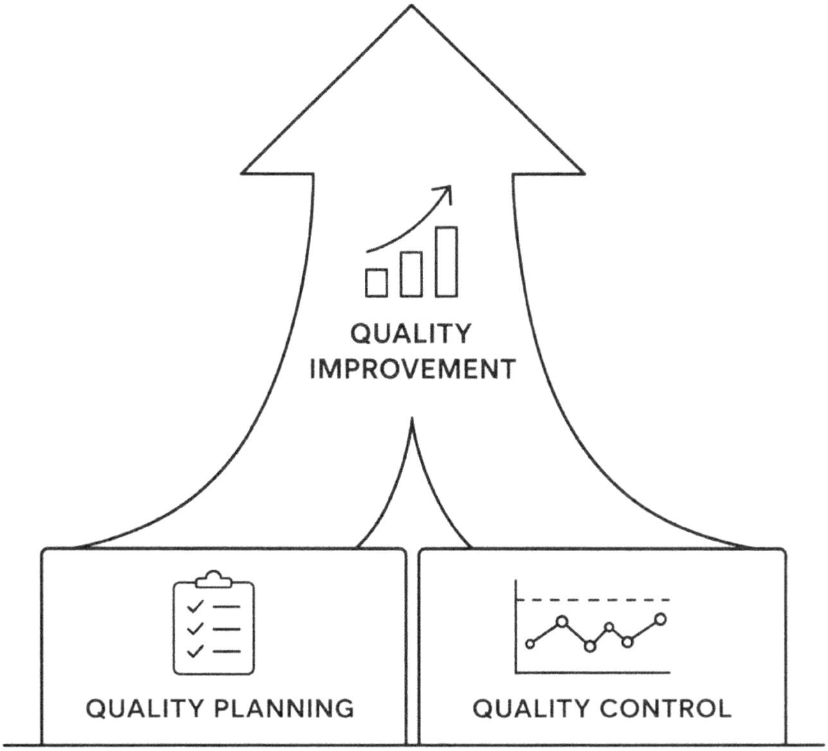

Figure 9.1 Continual improvement: the third part of Juran's trilogy.

Organising improvement as projects brings discipline and prevents improvement efforts from becoming vague ambitions or endless discussions. It also creates visibility and accountability. Improvement projects can be prioritised, resourced, and evaluated. In an increasingly project-oriented world, where change is implemented through temporary and focused efforts, this approach is becoming increasingly relevant.

Continuous improvement extends throughout the entire value chain. Suppliers, partners, and customers may all play a role in improvement initiatives. By engaging employees at all levels, organisations create ownership and commitment. Improvement is a shared responsibility, and in organisations with a strong quality culture, there is a genuine willingness to evolve. It is natural to look for better ways to perform tasks, simplify procedures, and reduce waste.

9.2 From reactive to proactive and predictive management

Improvement begins with understanding. In the previous chapter, we discussed measurement and control and distinguished between leading and lagging indicators. Lagging indicators tell us what has already happened and are useful for evaluations and comparisons. Leading indicators provide information during the process and enable intervention before undesirable results occur.

An organisation that relies primarily on lagging indicators tends to operate reactively. Problems are detected after they have materialised, corrective actions are taken, reports are written, and lessons are discussed. This is important, but rarely sufficient for sustained improvement.

A more mature approach is to be proactive. Being proactive calls for the systematic collection of information about what is happening in the operation while it is happening. Data must be gathered in real time or close to real time and transformed into useful information. Collecting technical data is not always possible, and in some contexts, it is not practical. However, the principle remains valid in most settings, and stability depends on access to relevant and timely information.

Consider a nursing home where health professionals work in shifts to care for patients. A nurse entering the night shift must be given an overview of the condition and status of the patients from the nurse responsible for the evening shift. Without such a structured transfer of information, there is a real risk that important details will be lost and that something serious may occur. In this context, being proactive may also include reporting incidents that nearly occurred but were avoided. Such "near misses" can provide valuable insights into weaknesses in routines or communication. Information may be collected through written reports or digital systems, but the essential point is that it is collected, shared, and used. At the same time, care must be taken not to create unnecessary administrative burden. Data should not be collected unless there is a clear purpose and willingness to use the data. Excessive reporting undermines motivation and weakens the system.

However, the situation is different in technical systems. Production processes, service platforms, and digital systems can often generate data automatically. Measurements can be taken continuously and converted into meaningful indicators through simple or automated calculations. Trends can be analysed, and key performance indicators can be monitored in real time. When used properly, such systems allow management to detect small deviations early and adjust before problems escalate.

A useful example from project management is the earned value method. This is a simple way of comparing what was planned, what has actually been achieved and what it has cost at a given point in time. By combining the original project budget, the assessed progress of the project, and the actual cost incurred, it is possible to calculate indicators that show whether the project is over budget, under budget or on track. The calculations are straightforward, but they require discipline, reliable data, and some competence in interpretation. The value of this method lies in its early warning capability. If deviations in cost or schedule are detected early, corrective actions can be taken before the situation becomes critical.

The transition from reactive to proactive management, and in some cases, predictive management, depends on access to relevant data, the ability to transform data into information, and the willingness to act on early signals. In the age of digitalisation, big data, and artificial intelligence, these possibilities are expanding rapidly. Advanced analytical tools can identify patterns that are difficult to detect manually and support decision-making in increasingly complex systems. The fundamental principle is to collect meaningful information, interpret it carefully, and use it to prevent problems rather than merely reacting to them.

Continuous improvement is not only about correcting errors but also about anticipating and preventing them. Practical methods and tools are required to achieve this. Several structured approaches are available to support data collection, analysis and disciplined

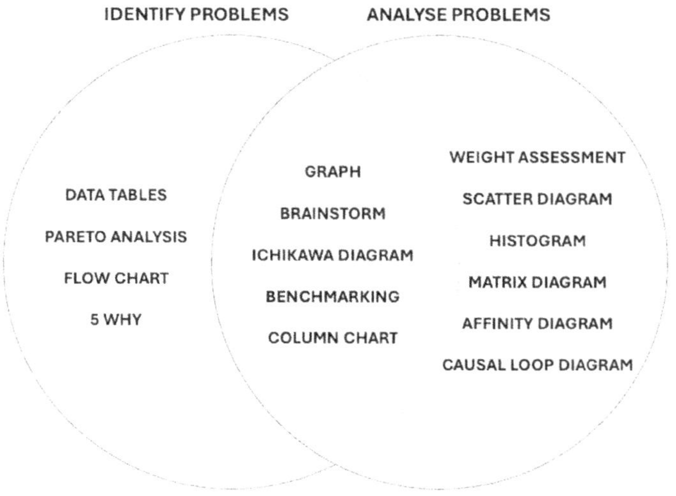

Figure 9.2 An overview of tools for identifying and analysing problems in improvement work.

problem-solving. Some of these are simple and intuitive, while others require more analytical competence. Together, these form a practical toolkit for systematic improvement.

Some of these tools are particularly useful for identifying that a problem exists. Others are more suitable for analysing the causes of a problem. Many tools can serve both purposes depending on their application. Figure 9.2 provides a simple overview of the tools presented in the following sections and illustrates how they can support different phases of improvement work.

9.3 Collecting and visualising data—identifying problems in practice

Improvement opportunities emerge from many different sources within and around an organisation. Internal audits may reveal weaknesses in procedures and support internal benchmarking between departments. Management reviews may highlight trends in performance. Customer complaints indicate dissatisfaction or unmet expectations. Employees may observe inefficiencies in their daily work. Improvement groups or quality circles may suggest ideas for improvement. Benchmarking, internally between departments or externally against competitors, may expose performance gaps in the organisation. Nonconformities may indicate specific variations in a process. Rising quality costs, excessive waste, and declining efficiency may indicate deeper structural issues. New requirements from regulators or customers may demand adjustments. In other cases, management may deliberately aim to increase process capability or reduce variations.

The need for improvement is sometimes obvious. A deviation has occurred, a requirement has not been fulfilled, or a process is clearly inefficient. In such situations, corrective actions can be defined relatively quickly. In other cases, however, the situation is less clear. A problem is perceived, but its nature is not fully understood. Complaints may be increasing, costs may be rising or employees may express frustration, yet there is no clear agreement about the root of the problem. In such situations, disciplined data collection is essential. Before solutions are proposed, the organisation must understand what is actually happening.

Problems in printing A4 sheets				
	Pages missing	Unclear	Wrong sequence	Total
Printer blocked				
Paperweight				
Humidity				
Toner				
Network				
Other			Total	

Figure 9.3 Example of a data sheet to identify the causes of problems.

A simple and effective starting point is the use of structured datasheets. These are used to record observations consistently so that patterns can be identified later. Instead of relying on impressions or anecdotal evidence, employees record relevant information whenever a deviation occurs. Over time, the collected observations provide a factual basis for discussion and analysis. For example, suppose that a computer printer in an office frequently malfunctions. The staff experienced delays and irritation, but the cause was unclear. Instead of guessing, a simple datasheet can be designed.

Each time the printer fails, employees record the time, type of malfunction, and possible contributing factors, such as paper type or software used. After some time, patterns may emerge in the data. It may turn out that most failures occur when a specific type of document is printed or during peak usage hours. What previously seemed random may now be revealed as a pattern.

Problems with lunch delivery	Week 1	Week 2	Week 3	Week 4	Total
1. Delivered too late					
2. Delivered too early					
3. Too cold					
4. Bad taste					
5. Too large portions					
6. Too small portions					
7. Too many portions					
8. Too few portions					
Total					

Figure 9.4 Example of a data sheet to seek the causes of problems by monitoring changes over time.

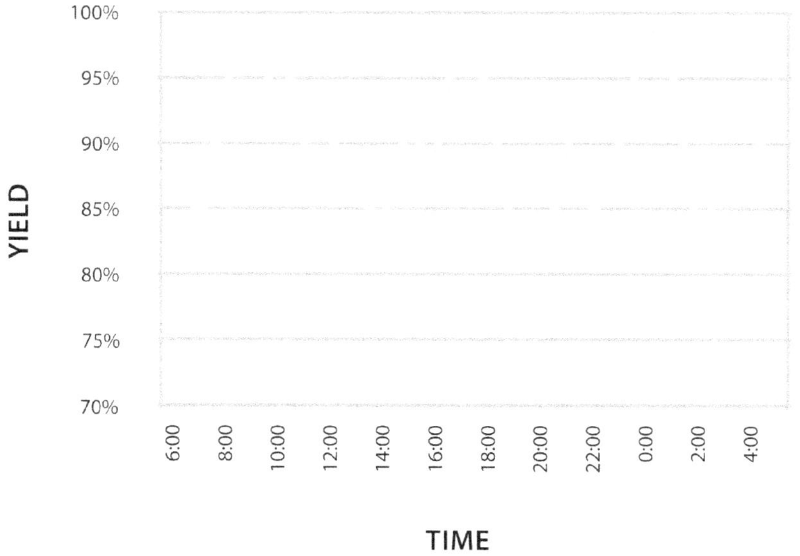

Figure 9.5 A time-series graph showing the yield as a function of time.

Another example is a company that purchases lunch from external suppliers for its employees. Complaints have been made, but the nature of the problem is unclear. Is the food too cold? Is the delivery late? Is the quantity insufficient?

A simple registration sheet can be created in which employees can record specific complaints over a defined period. After collecting data for a few weeks, it may become evident that the main issue is that meals arrive late on certain days. The organisation now has a factual basis for discussion with the supplier.

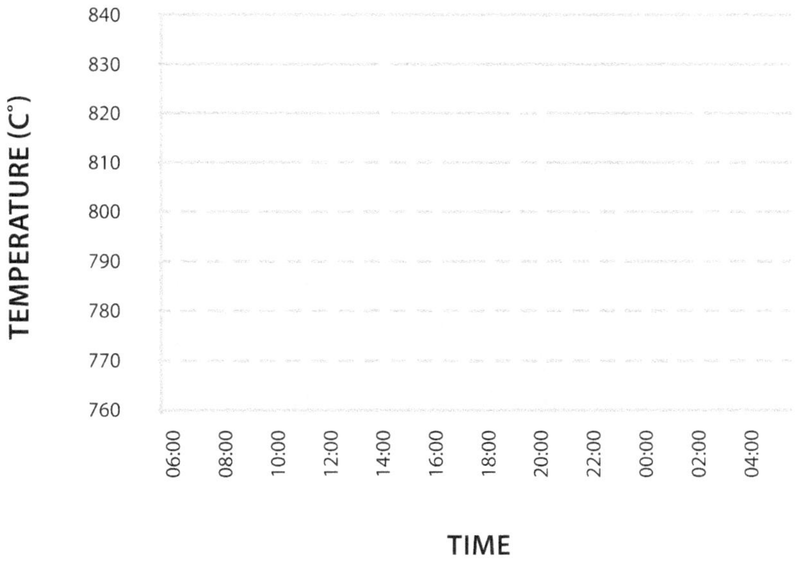

Figure 9.6 A column chart that shows how the average temperature changes with time.

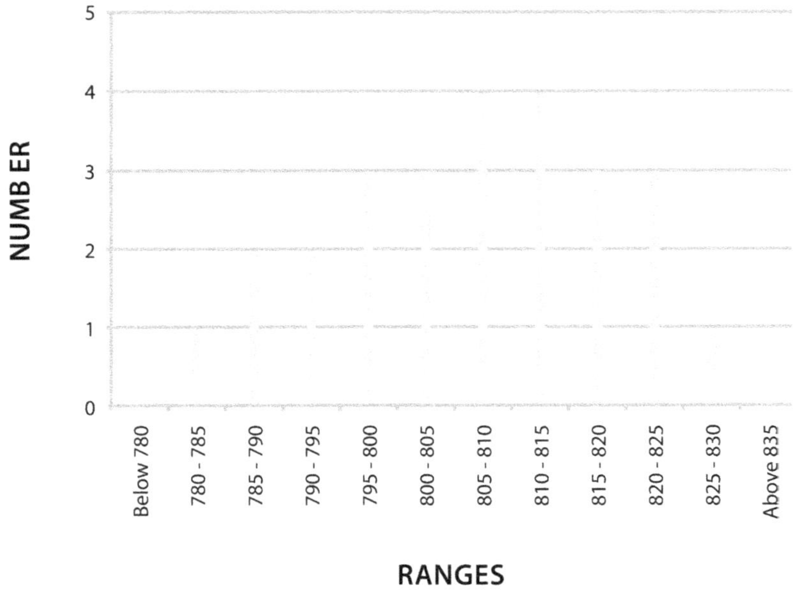

Figure 9.7 A histogram showing the distribution of the data.

Monitoring problems over time is also a powerful approach. Trends can be identified by counting the number of incidents or deviations each week and plotting them in a simple graph. A time-series graph may show whether the situation is improving, deteriorating, or fluctuating. Figure 9.5.5 shows how the yield in an industrial process change over a period of 24 h. Visualising data in this way often makes it easier for everyone to agree on the existence and magnitude of a problem.

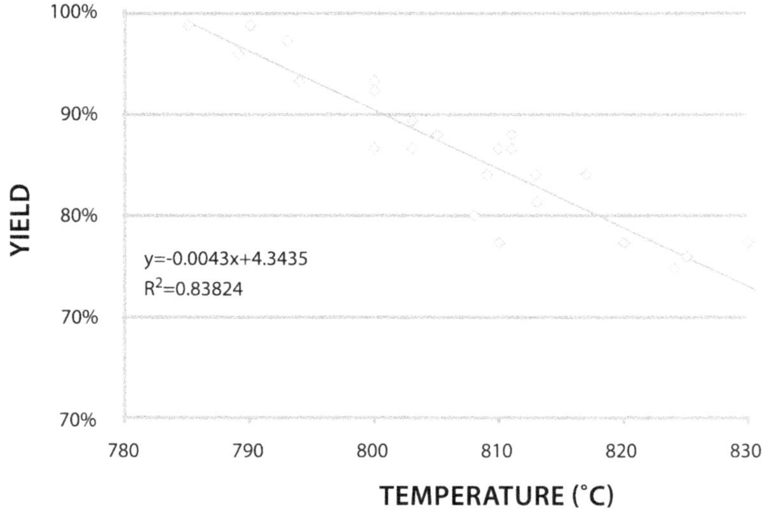

Figure 9.8 An *X–Y* chart to show the correlation between yield and temperature.

Column charts are used to sort and present information in an organised manner. They are particularly useful for visualising data collected over time or for comparing the frequency of different categories. By presenting data in a simple visual format, column charts make it easier to identify patterns and discuss improvement priorities. Figure 9.6 shows how the average temperature in an industrial process change over a period of 24 h.

Histograms are particularly useful when dealing with measurable variables, such as temperature, processing time, and response time. Consider a production process in which the temperature is measured every hour. The distribution becomes visible by grouping the measurements into intervals and plotting the frequency of each interval. The histogram may reveal that the process frequently operates near the upper limit of the acceptable temperature, which may increase the risk of defects. Figure 9.7 is an example of such a histogram, and it is based on the same data as the column chart shown in the previous figure.

Scatter diagrams or X–Y charts are used when there is reason to believe that two variables are related. For example, the efficiency of a production process may be influenced by temperature. By plotting the efficiency against temperature, it is possible to determine whether higher temperatures are associated with lower yields. If a clear pattern emerges, further investigations can focus on stabilising the temperature to improve efficiency. Figure 9.8 is an X-Y chart based on the same data as the last figures. In this case, there was a significant correlation between yield and temperature. This relationship is evident when looking at the graph and is confirmed by modelling the relationship as a straight line and calculating the correlation coefficient.

Pareto analysis (as shown in Figure 9.9) helps prioritise improvement efforts by highlighting the most significant problems. Suppose that several types of defects occur during a manufacturing process. When these are arranged in descending order according to frequency, it often becomes clear that a small number of defect types account for a large proportion of problems. This is sometimes referred to as the 20/80 principle,

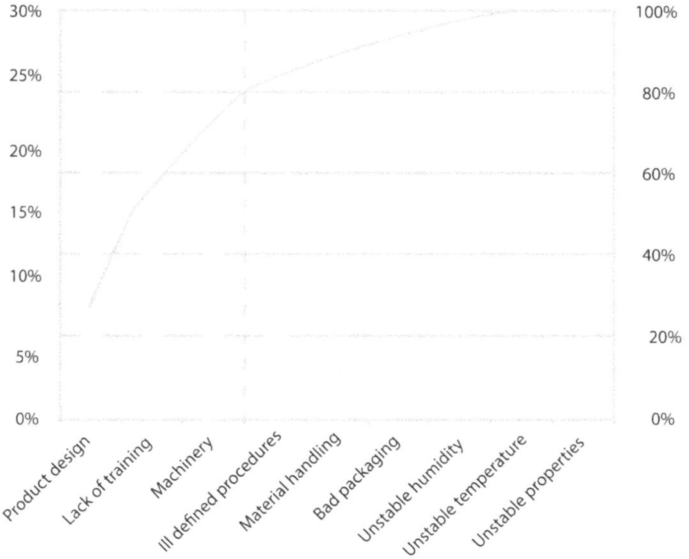

Figure 9.9 Example of Pareto analysis to identify the most important problems in a production process.

where roughly 20% of the causes explain approximately 80% of the effects, although the exact proportions may vary. Rather than spreading resources evenly across all issues, the organisation can first focus on those that have the greatest impact.

These tools are simple but powerful when used with discipline. Their purpose is to make problems visible and measurable. A graph does not explain why something happens, but it clarifies what is happening. This creates a shared understanding of reality.

Once patterns and deviations are identified and visualised, the next step is to ask why they occur. In the following section, we move from observation to explanation and examine structured methods for identifying root causes.

9.4 Diagnosing root causes

Once a problem has been identified and described with the help of data, the next question is why does it occur? It is tempting to move directly to solutions, and this is where many improvement efforts fail. People rely on intuition, quick judgements, or majority opinions. The result may be that symptoms are treated while the real cause remains untouched. Therefore, root cause analysis is a crucial discipline for improvement work.

The tools discussed above are powerful when problems can be clearly observed and described. In many organisational situations, however, the challenge is not only to analyse quantitative data but also to make sense of qualitative information—opinions, experiences, perceptions, and interdependencies. In such cases, additional tools may prove useful.

A simple and widely used method is brainstorming sessions (Figure 9.10). When applied correctly, brainstorming allows participants to suggest possible causes freely without criticism. All ideas are recorded. Only after the ideas have been collected does the group discuss, clarify, and categorise them. The strength of brainstorming lies in its ability to draw on employees' collective experience. Often, people working close to the process have valuable insights that would otherwise remain unarticulated.

To structure such discussions further, the Ishikawa diagram, named after the Japanese quality pioneer Kaoru Ishikawa and known as the fishbone diagram, is often used. The problem is briefly defined on the right side of a large sheet of paper. From this, a horizontal line is drawn, and possible categories of causes are identified. A common classification includes categories such as people, equipment, processes, and materials.

Let us return to the example of lunches arriving too cold to eat (Figure 9.12). The problem, "Lunches are too cold when received" is written at the top of the diagram. Under "process," possible causes might include unclear delivery procedures. Under "people," there may be insufficient training of the delivery staff. Under "equipment," there may be inadequate transport containers. Under "materials," perhaps inappropriate packaging is mentioned. As the diagram develops, certain causes may appear repeatedly in the diagram. For example, the lack of defined procedures or training may be mentioned in several branches. This repetition often indicates deeper structural issues.

After listing the possible explanations, recurring themes emerged. In this case, the lack of procedures and follow-up appeared more than once. A closely related cause, lack of training, also appears repeatedly. The root causes of the problem have now been identified, and attention can be shifted towards improving procedures and ensuring proper training.

The Five Whys is another method for seeking the underlying causes of a problem. This approach was originally presented by the founder of Toyota, Sakichi Toyoda, and forms

Figure 9.10 Brainstorming session.

part of the Toyota production system. The method is iterative and encourages participants to continue asking "why?" until the fundamental cause has been identified.

The simplest way to do this is to write the problem clearly on a sheet of paper. The group then asks why this problem occurs. Each answer leads to a new "why?" question. This process continues until the root cause is revealed. The aim is to move beyond immediate symptoms.

Figure 9.13 shows a simplified example applied to incorrect invoices. In that case, four iterations were sufficient to reveal that the real problem was not merely individual error but the lack of formal procedures and an organisational structure that had not adapted to increased activity. Five is not a sacred number; what matters is persistence and logical consistency.

When several possible causes or solutions have been identified, a weight assessment or simple prioritisation exercise can be useful. Participants may be asked to assign points to alternatives. This allows all voices to be heard and helps the group reach a consensus on what to address first. This structured approach reduces the influence of hierarchy and personal dominance in discussions.

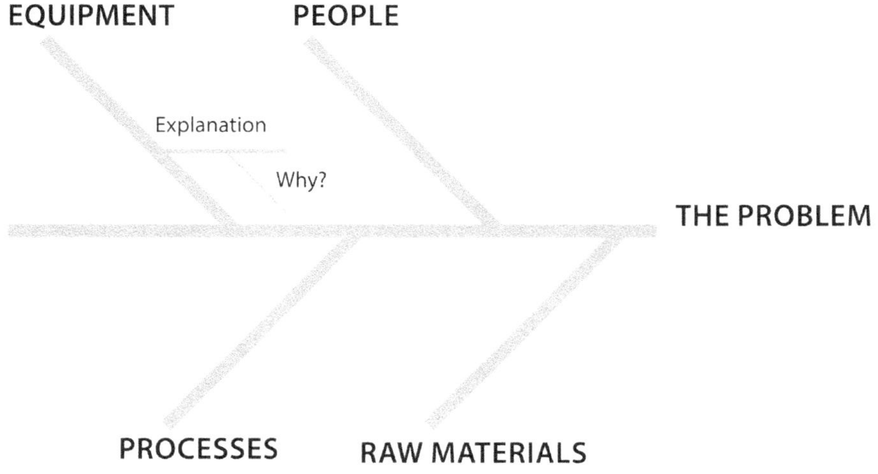

Figure 9.11 Fishbone diagram to search for the root causes of problems.

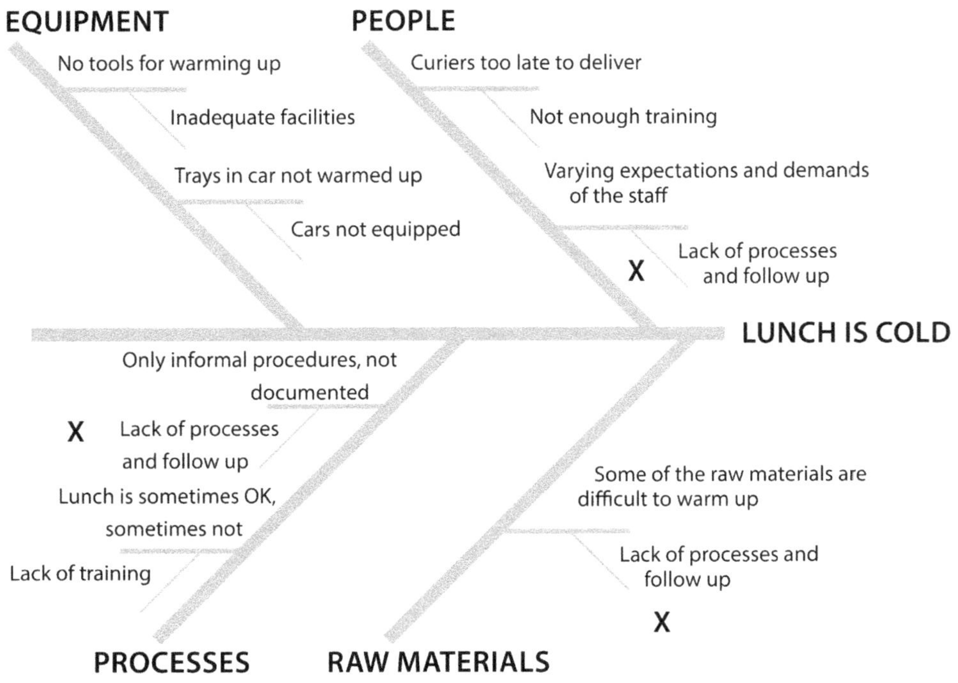

Figure 9.12 Example of a fishbone diagram to search for causes of the problem that lunches are too cold when received.

In more complex situations, where many ideas and perceptions are involved, an affinity diagram can be useful. After brainstorming, ideas are written on separate notes and randomly placed on a wall. The group then clustered similar ideas together without initially defining the categories. Gradually, themes emerge—for example, communication

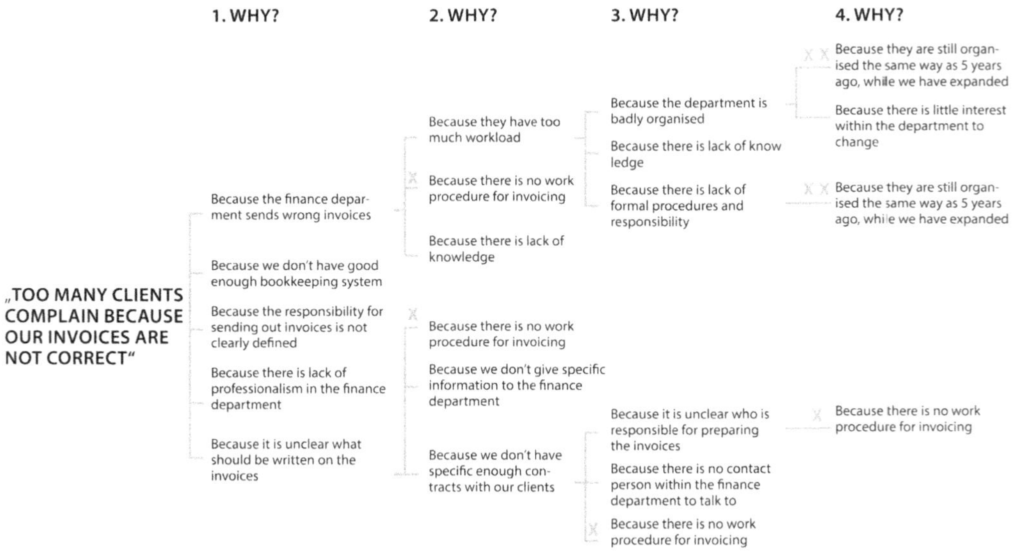

Figure 9.13 The 5 why method applied to look for root causes of a problem with wrong invoices.

Figure 9.14 Creating an affinity diagram.

issues, service delays, or product design problems. Once clusters are formed, they are labelled. The affinity diagram helps transform a collection of scattered thoughts into a structured insight.

A matrix diagram is useful when the relationships between two or more sets of factors need to be clarified. For example, a company may want to understand how different departments contribute to specific quality objectives. Departments are listed on one axis, objectives on the other, and the connections are marked. Overlaps and gaps quickly become visible. In project management, the well-known RACI matrix, which clarifies who is Responsible, Accountable, Consulted and Informed, is a practical example. Matrix diagrams are particularly helpful when coordination and role clarity are part of the issue.

Some problems are not linear but systemic. The causes and effects influence each other through feedback loops. In such cases, a causal loop diagram, sometimes called a relations diagram, can help to visualise dynamic relationships. For example, an increased

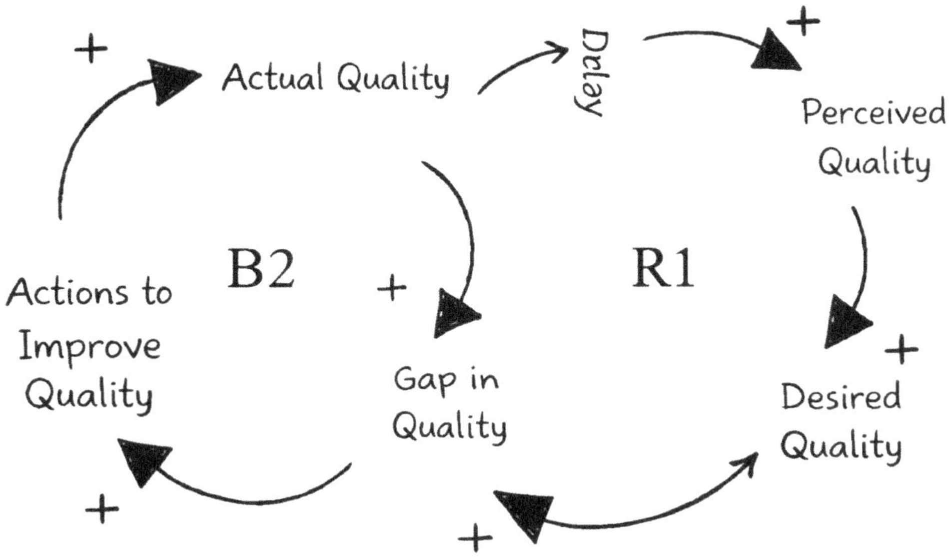

Figure 9.15 A simple causal loop diagram.

workload may lead to stress, which may increase errors, which create rework, which further increases the workload. This is a reinforcing (R) loop. By mapping such relationships with arrows indicating the direction and type of influence, an organisation gains a broader understanding of system behaviour. Causal loop diagrams are especially valuable in complex environments, where improvement in one area may unintentionally create problems elsewhere.

Root cause analysis requires a great deal of patience. The temptation to implement quick fixes is strong, especially when the pressure for results is high. However, acting without understanding often leads to recurring problems. Addressing the true cause leads to sustainable improvement.

Once the most likely causes have been identified and agreed upon, the organisation is ready to move from analysis to structured action. In the next section, we examine how improvement initiatives can be organised as disciplined processes, ensuring that solutions are not only identified, but also implemented and sustained.

9.5 A structured process for improvement—Kume's seven steps

Improvement requires a structure. Identifying problems and analysing their causes is important, but it is crucial to implement solutions in a disciplined and sustainable manner. One practical framework is the seven-step process (figure 9.16) presented by Hitoshi Kume (Kume, 1987)[1]. This is, in essence, a project life cycle model for organisational improvement work. It builds directly on the logic of the plan–do–check–act (PDCA) cycle. Planning corresponds to defining and analysing the problem. Doing relates to the implementation of solutions. The checking process involves performance verification.

Figure 9.16 The seven-step process.

Acting includes standardisation and learning. In this way, the concepts introduced earlier now come together in a concrete method for disciplined improvement.

Step 1—Definition of the project

The first step is to clearly define the project. The importance of the problem is described, and the scope is agreed upon. Some of the questions to be answered are as follows:

- What is the story behind this problem?
- What are the unfortunate consequences of the problem?
- How important is this project?
- What is the problem all about?
- What is the scope of this project?
- What resources are needed to solve it?

An organisational structure is created for the project, and a time plan and a plan for resources and costs are defined. A project manager is appointed, meetings are scheduled, the time limit for the project is set, and performance indicators are defined. It is common that more information is needed in this first step, and there is a risk of jumping to conclusions. The definition of the scope can be tricky, and there is a tendency to have too wide a perspective and the desire to solve all problems in the first step. The risk is that the overview of the real problem is lost.

Step 2—Consider and understand the problem

In this step, the team develops a shared understanding of what the problem entails. Processes may be mapped, stakeholders identified, and additional data collected. The tools presented in the earlier sections are particularly relevant here. For instance, it may be useful to draw a flowchart that represents a common understanding of the process under investigation.

At the end of this step, the team agrees on an overview of the problem, which includes a description and perhaps some numerical data regarding the problem. A frequent mistake is attempting to solve a problem before it is properly understood. Without a factual and shared understanding, subsequent steps may be based on weak assumptions.

Step 3—Identifying the root causes

Based on the collected data and structured analysis, possible causes are identified and tested. It is important to distinguish between symptoms and their root causes. Tools such as Ishikawa diagrams, the Five Whys technique, affinity diagrams, matrix diagrams, or causal loop diagrams may support this analysis.

The outcome of this step is the group's common view of the most likely cause of the issue under investigation. A frequent challenge at this stage is that people tend to be hasty and move directly to implementing solutions without proper planning. It is also common not to use the available data fully but instead to rely on intuition or follow the view of the majority. Finally, it can be difficult to maintain focus. There is often a tendency to search for the causes of many issues simultaneously rather than concentrating on the most important one at hand.

Step 4—Finding and implementing a solution

Once the root causes are identified, a solution can be proposed. The objective is to eliminate the causes rather than treat the symptoms. Support from relevant stakeholders is essential, particularly when changes affect routines or responsibilities.

In this step, support for the proposed solution is developed, and care must be taken to ensure that it does not create negative side effects or lead to new issues. An implementation plan is then prepared, and the solution is put into practice. A common challenge at this stage is insufficient support from key stakeholders, particularly managers, even though their commitment is often essential for implementing changes that may be demanding or uncomfortable for them.

Step 5—Verify performance

This step reflects the "check" phase of the PDCA. After implementation, the results must be measured. A comparison with the baseline data is essential. Have the defined objectives been achieved? Have any unintended side effects occurred?

The outcome of this step is that measurements are available, evaluated, and a decision is made on the next steps. Typical problems at this stage are neglecting to firmly define a new process or solution and making it permanent. This is the most vulnerable part of the implementation process.

Step 6—Uphold

This step reflects the "act" phase of the PDCA. If improvements are achieved, they must be anchored in the management system. The procedures may need to be updated. Training may be required for this. The responsibilities for monitoring must be defined. This step ensures that improvements become part of normal operations rather than temporary initiatives.

A necessary training plan is required, and the new procedures must be formally approved in the QMS. Responsibility must be defined, in terms of measurements and registrations, if this is the case. The result of this step is typically a new or updated quality document, where the new procedure is described and responsibility is defined. The problem here is often a lack of communication; information is not flowing efficiently, and there is a lack of preparation and support for actions that have not been obtained prior to implementation. In addition, training is sometimes lacking, and measuring and monitoring are neglected.

Step 7—Drawing conclusions

Finally, the team reviews the entire process. What was learned? What worked well? What can be improved in future projects? Good results should be acknowledged, and participants should be recognised for their contributions. This step reinforces organisational learning and strengthens the capability for future improvement efforts.

Kume's seven-step process illustrates how continuous improvement can be organised in a disciplined and practical manner. It integrates analysis, implementation, and learning, bridging problem-solving and operational standardisation. The framework has broad applications in continuous improvement and structured problem-solving. It combines qualitative and quantitative data, encourages visual tools and team-based analysis, and is suitable for organisations of different sizes and sectors. Perhaps most importantly, the framework is simple and accessible to all. This makes it valuable for building a culture of improvement in the field.

Another well-known framework for improvement projects is DMAIC, which is closely associated with Six Sigma. DMAIC refers to the steps Define, Measure, Analyse, Improve, and Control. Like Kume's seven-step process, DMAIC provides a structured approach to identifying problems, analysing causes, implementing solutions and sustaining results. Six Sigma is particularly associated with reducing variation and improving process capability through disciplined use of data and statistical methods. In practice, the underlying logic of DMAIC and Kume's framework is very similar, although Six Sigma is often more formalised and data-intensive.

In the following section, we turn to organisational approaches that support sustained improvement in everyday operations.

9.6 Lean thinking and visual management—making improvement visible

Structured improvement projects are important. However, continuous improvement cannot be solely reliant on occasional projects. It must also be embedded in the daily operations. Lean thinking offers such a perspective. Lean was introduced in Chapter 2 as part of the evolution of quality management. We now take a closer look at two core ideas from Lean.

Looking for and elminating waste

Lean management is an ideology that focuses on delivering maximum value to customers while minimising waste. The principle is simple; time, resources, and effort should not be used unless they contribute to value. In this sense, lean thinking is closely aligned with the core ideas of quality management. Lean thinking encourages employees to critically observe their daily work: Where is value created? Where does time disappear? Where are the errors generated? Where can the processes be simplified? This way of thinking is closely connected to the Japanese concept of Kaizen, which can be translated as "continuous improvement." The idea is that improvement should not depend only on large initiatives or technological breakthroughs. Small, ongoing improvements made by employees in their everyday work can collectively create a major long-term impact. Improvement then becomes part of organisational culture rather than an occasional activity.

In many organisations, Lean and Six Sigma are used together in what is often referred to as Lean Six Sigma. In simplified terms, Lean focuses on flow, waste reduction and value creation, while Six Sigma emphasises the reduction of variation and disciplined use of data. The combination aims to improve both efficiency and consistency in organisational processes.

In The Toyota Way Fieldbook, Liker and Meier (2006)[2] describe eight types of waste that commonly occur in organisations. These categories are highly practical and can be recognised in almost any business.

1. Overproduction. Producing more than needed or earlier than needed. For example, printing large quantities of reports that are never read or manufacturing products before confirmed orders are received. Problems are created in terms of storage, transport, and coordination, which calls for labour that could otherwise be saved.

2. Unnecessary waiting time. Employees waiting for materials, decisions, or information. In an office environment, this may occur when a case cannot proceed because of a missing signature.
3. Unnecessary transportation or handling. Moving products, documents, or information unnecessarily. Even digital systems can create "transport waste" when information must be re-entered multiple times.
4. Producing too much. Unnecessary work is performed in the process, for example, due to the use of incorrect tools or because of the manufacturing process. This also includes manufacturing a product to a higher specification or providing a higher-quality service than requested.
5. Unnecessary supplies. Accumulating materials, unfinished products, or tasks in queues. In service organisations, this may take the form of backlogs. This results in long waiting times, damage, unnecessary shipping costs, and delays.
6. Unnecessary movement. All movements that employees must make in the production process that are not intended to increase the value of the product. This also includes when employees need to move from one place to another during production without increasing the value of a product.
7. Defects and rework. Producing something that does not meet the requirements. Incorrect invoices, defective components or errors in documentation all lead to rework. When a defective product is produced, it is recycled or discarded; this is waste.
8. Unused employee creativity. Failing to utilise employees' knowledge and ideas. This may be one of the most costly forms of waste in knowledge-based organisations.

These categories demonstrate that lean is not limited to the manufacturing sector. Waste can occur in hospitals, universities, consulting firms, public administration, and all kinds of organisations. The purpose of identifying waste is to improve the systems.

Visual management: Clarity instead of bureaucracy

Lean thinking is closely related to visual management. The core idea is to make information visible and accessible so that actions become easier. Quality management is sometimes criticised for being overly focused on documentation and reporting. Visual management challenges this perception. Instead of long written reports, improvement work can be summarised clearly and concisely.

A well-known example is the A3 report for coordinating improvement projects. An A3 sheet can contain the following:

- The problem definition
- Analysis of causes
- Proposed actions
- Implementation plan
- Results
- Next steps

Everything fits on a single sheet as shown in Figure 9.17. If the problem cannot be explained on a single A3 sheet, it is likely not fully understood. The A3 approach reduces unnecessary reporting while ensuring that essential information is documented properly.

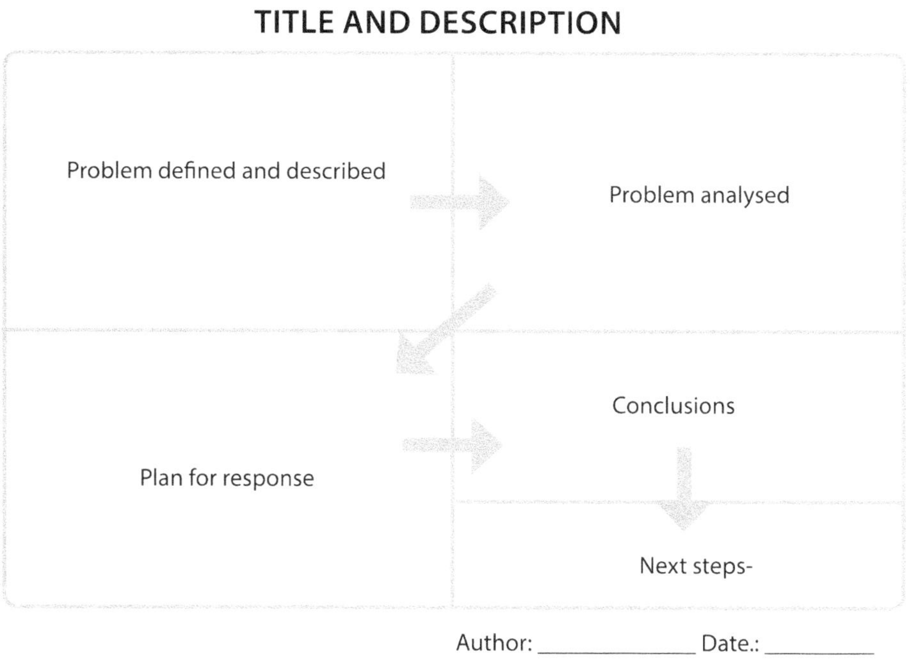

Figure 9.17 A3 report for coordinating improvement projects.

The purpose of the A3 sheet is to avoid needless reporting and save time on unnecessary registration while ensuring that everything that matters is registered.

This idea can be expanded beyond a single project. An improvement board can be placed in a visible area where ongoing and proposed improvement activities are displayed to encourage participation. Figure 9.18 gives an example. At the top, measurable quality objectives may be shown to maintain a clear focus.

Improvement projects are defined by a descriptive title, a short list of main actions, the responsible person, measurable goals, and the expected completion date. This information can be written on a small Post-it® note and placed on the board. The area titled "project in progress" shows the status of ongoing projects, and the notes are moved from left to right as the project progresses through defined stages. When a project is completed, it disappears from the active area, and one descriptive line is added to the "concluded projects" section.

A well-designed board ensures that everyone has an overview of what is going on and an understanding of what their colleagues are working on. This means that the approach to improvement projects is consistent and clear for everyone. If this process needs to be adjusted, the layout of the board can easily be changed. This can happen in parallel with its use; the process develops and alters, indicators may change, and measurements are obvious.

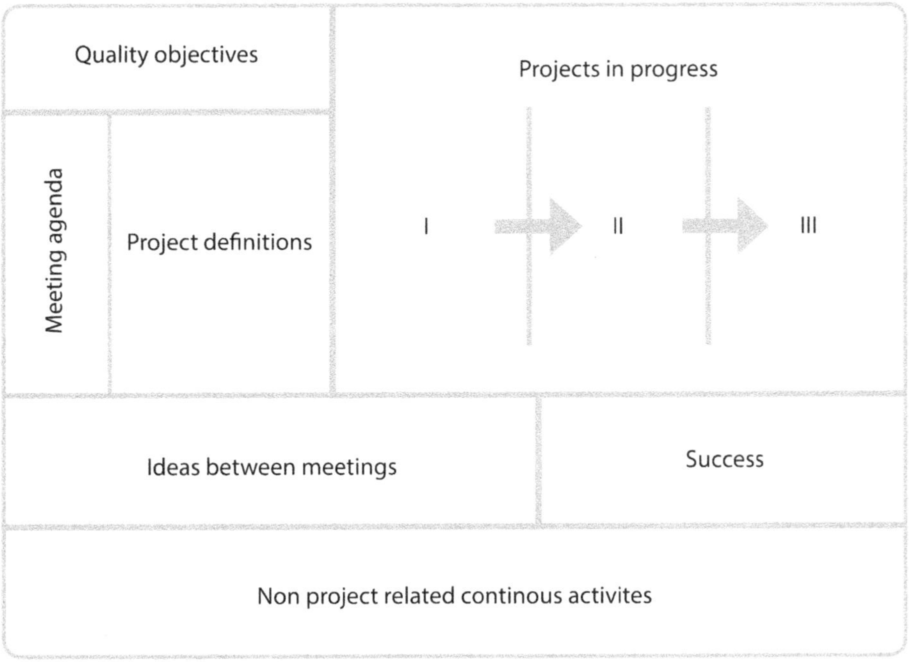

Figure 9.18 A board to handle all improvement projects in a small company.

Meetings can then take place in front of the board. Preparation is minimal. Those responsible update the status of their projects and briefly explain their progress. The board ensures that:

- Everyone has an overview of what is happening.
- Goals are visible.
- Responsibilities are clear.
- Attention is directed towards obstacles and next steps.

When properly used, visual management strengthens communication between managers and employees, reduces unnecessary reporting, shortens meetings, and supports focused follow-up. Thus, lean thinking and visual management demonstrate that quality management does not have to become bureaucratic. When implemented thoughtfully, it simplifies work and clarifies priorities.

9.7 External audits and certification—an outsider's perspective

Internal audits were previously discussed as an important mechanism for professional autonomy within a quality management system. They provide a structured way for the organisation to review its processes and verify that procedures are followed and that the system functions as intended.

External audits provide a different perspective and can take many forms. A large client may audit a supplier before entering into a contract. Authorities may conduct inspections to verify compliance with the regulations. Key stakeholders may request documentation

Figure 9.19 External auditor reviewing the QMS, including internal audits.

or perform reviews before entering a cooperation. In all these cases, an independent party examines the organisation's management systems and practices.

One specific and widely known form of external audit is certification according to international management standards such as ISO 9001 or ISO 14001. In this case, a certification body assesses whether the organisation's management system fulfils the requirements of the relevant standard and whether it is effectively implemented in practice. Independence is central; the external auditor is not part of the organisation.

The external auditor assesses the entire management system, as indicated in Figure 9.19. A crucial part of this assessment is the organisation's internal audit process. If internal audits are weak or merely formal, the credibility of the entire system is undermined. Active and disciplined internal audits therefore form an essential foundation for successful certification.

Why pursue certification?

The decision to pursue certification is strategic. It requires resources, preparation and long-term commitment. There are several reasons why an organisation may choose this path. Certification may be required by stakeholders in an organisation's environment. Regulators may require formal management systems and large customers may require their suppliers to operate certified systems. In international markets, certification may serve as a prerequisite for participating in tenders or partnerships.

In other cases, certification is pursued to verify the management system and obtain independent confirmation that the system meets recognised standards. This external confirmation can strengthen credibility and trust among customers, owners, and partners.

Certification can also serve as a powerful internal driver. Declaring the intention to achieve certification by a certain date sends a clear message to the entire organisation. It creates focus, mobilises resources, and unites employees around a concrete goal. Certification can serve as a milestone in the broader development of quality management.

It is important that certification is pursued for the right reasons. If the primary objective is merely to obtain a certificate for marketing purposes, the result may be a superficial system designed to pass audits. Documentation may increase without improving the performance and employees may experience the management system as a bureaucracy rather than as support.

However, when certification is approached as part of a genuine improvement strategy, it can create value. An external audit provides an outsider perspective. Auditors typically have experience in many organisations and industries. They may question established routines, identify weaknesses, and highlight risks that internal personnel no longer see. Simple questions such as "How do you know this process is effective?" or "Is this procedure consistently followed?" can stimulate valuable reflections.

Certification should therefore be understood as a milestone, not a destination. Achieving certification confirms that a system meets the defined requirements at a given point in time. This does not guarantee continuous improvement. Maintaining certification requires ongoing internal audits, management reviews, and systematic follow-up.

Consultant and certification body—two different roles

It is important to clearly distinguish between the roles of a consultant and a certification body. A consultant may assist an organisation in designing and implementing its management system and is hired for expertise and experience.

A certification body plays a fundamentally different role. It must remain neutral and independent. The certification body audits the system but does not design it. In practice, the certification body is also evaluating the results of the consultant's work. Therefore, a certification body cannot provide consulting services to the same organisation it certifies. Independence is the cornerstone of credibility.

This distinction reinforces a key principle already discussed: those who design or operate a system should not be the ones who audit it.

Accreditation and conformity assessment

In the context of certification, the concept of accreditation must be introduced. Accreditation can be defined as the formal approval by a competent authority of an entity's ability to perform specific conformity assessment tasks. Conformity assessment refers to the evaluation of whether a product, process, system, or organisation meets defined requirements.

When a certification body certifies a company's QMS, accreditation assures that the certification body is competent to carry out this evaluation. Accreditation bodies operate at the national or international level and assess certification bodies against defined criteria.

Two forms of accreditation can be distinguished. In some areas, accreditation is legally required, for example, in activities such as calibration or safety inspections. In other cases, accreditation is voluntary but pursued to strengthen credibility and competitive positions.

Choosing a certification body

If an organisation intends to seek certification, it is advisable to proceed systematically. Certification bodies operating in the relevant area should be reviewed, their reputation examined and their accreditation verified. Direct contact should be established to discuss availability, scope, costs and practical arrangements.

Some certification bodies offer preliminary audits before formal certification. Such assessments can provide valuable insight into the organisation's readiness and allow time for necessary adjustments.

External audits also strengthen trust. Customers, partners and other stakeholders gain confidence when an independent body evaluates the system. For smaller organisations in particular, certification may support access to new markets.

At the same time, certification introduces discipline. Once an organisation commits to a management standard, it commits to maintaining and developing its system. Changes in processes, strategies or structures must be reflected in updated procedures and records.

In this broader context, external audits are part of the continuous quest to become better and better.

9.8 Developing beyond compliance—guidance for sustained success

Certification according to ISO 9001 confirms that a management system fulfils the defined requirements. It demonstrates conformity. However, conformity is not the same as long-term success. Organisations that genuinely wish to develop further must ask a broader question: How can we maintain and strengthen our performance over time in a complex and changing environment?

Several approaches can be used to address this challenge. Excellence models, such as the EFQM, provide structured maturity frameworks. Additional ISO management standards, such as ISO 14001 for environmental management and ISO 45001 for occupational health and safety, expand the scope of organisational responsibility. Earlier in this book, we discussed the aligned framework of ISO management standards and how their common structure supports coherence.

Broadening the scope: Towards integrated management

As organisations mature, improvements often extend beyond quality alone. Environmental responsibility, occupational health and safety, information security, and other areas have become increasingly important. Rather than operating several parallel systems, many organisations choose to integrate these perspectives into a coherent management system. An integrated management system aims to reduce the duplication of processes and documentation while strengthening coherence. Shared processes for leadership, planning, risk management, internal audits, and management reviews can support multiple objectives simultaneously.

Earlier, we discussed the aligned structure of ISO management standards. This structural compatibility makes the integration feasible. When done thoughtfully, integration can strengthen strategic alignment and ensure that quality, environmental, and social considerations are addressed together rather than in isolation.

In the context of the quest to become better, integration represents a step towards maturity. It reflects systems thinking and recognises that organisations operate within

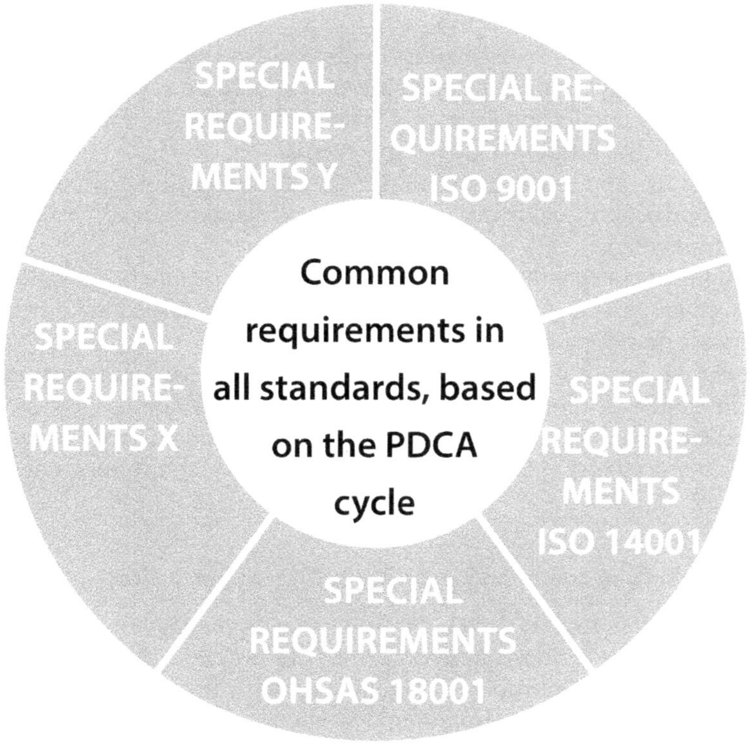

Figure 9.20 The requirements of the management standards are largely common and based on the PDCA cycle.

broader ecosystems and that different objectives must be balanced rather than optimised separately.

ISO 9004

ISO 9004 is a useful contribution to this broader landscape. Unlike ISO 9001, ISO 9004 is not a certification. This is a guidance document intended for organisations that seek sustained success. The focus is on organisational quality and long-term capability.

ISO 9004 defines the quality of an organisation as the degree to which its inherent characteristics fulfil the needs and expectations of its interested parties. Quality concerns the organisation as a whole, including its leadership, culture, strategy, processes, and relationships. A central concept is the identity of the organisation, reflecting the interaction among mission, vision, values, and culture. Sustainable success requires alignment of these elements. Declared intentions must be reflected in daily behaviours. Strategies must be supported by competence and culture.

Another important feature of ISO 9004 is its emphasis on maturity and self-assessment (SA). The standard includes a maturity model that enables organisations to evaluate their development from basic to best practice. This is not an external audit instrument but a reflective tool for internal learning. It helps organisations identify strengths, weaknesses and improvement priorities in a structured way.

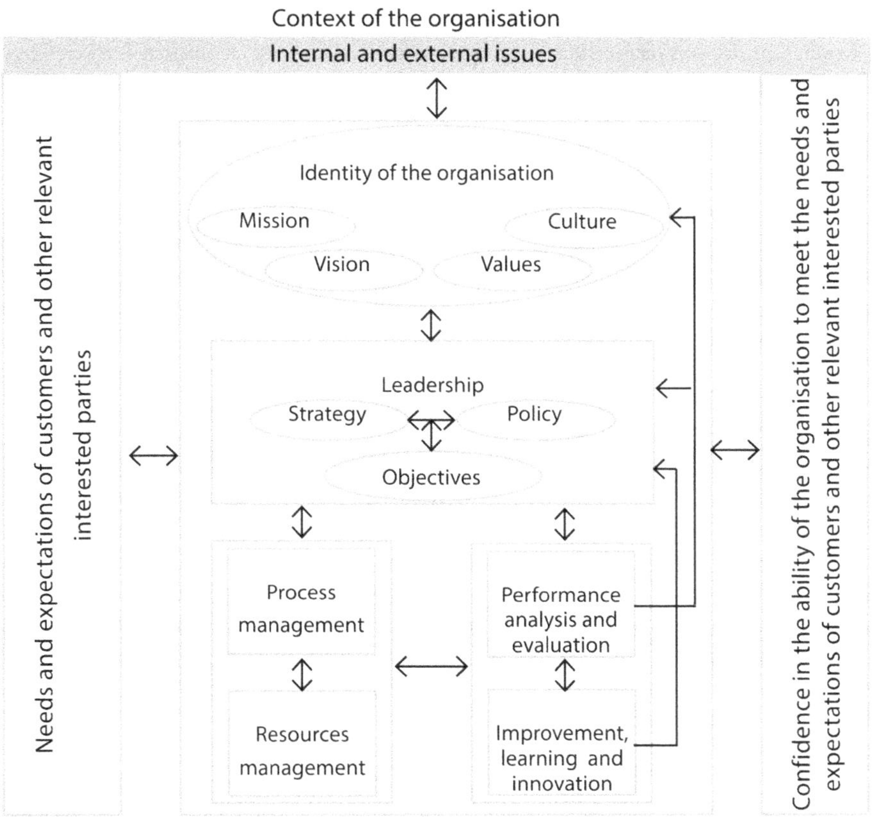

Figure 9.21 A graphical representation of the ISO 9004 standard.

For organisations that truly wish to improve, such self-assessment can be extremely valuable. It shifts the discussion from "are we compliant?" to "how capable are we?" This encourages long-term thinking and systematic development.

In many ways, this reflects the evolution of quality management. The field has moved from inspection and defect control to process management and customer focus, and further towards organisational resilience and sustainability. Although efficiency remains important, it is no longer sufficient. Organisations must also develop the capacity to learn, innovate, and adapt.

The message is clear; a management system should strengthen the organisation's capability. It should enhance the alignment between purpose, strategy, processes, and culture and support sustained success.

This broader perspective prepares us for final reflections. If quality management is fundamentally about the continuous quest to become better, what does that mean in a world shaped by sustainability challenges, digital transformation, artificial intelligence, and a growing reliance on projects to drive change?

9.9 From compliance to capability—purpose in an evolving world

In this book, quality management is presented as a way of thinking and acting. From discussions of quality culture and process understanding, through measurement, control, and structured improvement, to reflections on sustained success, one central idea has remained constant. Organisations must never be satisfied with the status quo. Continuous improvement is not an isolated activity. It is a mindset.

The quality trilogy introduced by Juran—planning, control, and improvement—remains highly relevant. Planning provides a direction. Control ensures stability. Improvements drive development. Without improvement, planning and control merely preserve the existing performance. We emphasise the importance of moving beyond compliance. A management system designed merely to satisfy external requirements will never unlock its full potential. Documentation is necessary but is not the goal. The goal is capability, which is the ability of the organisation to fulfil its purpose consistently and responsibly.

This distinction is becoming increasingly important in today's environment. Organisations operate within complex systems. Value chains evolve into value networks. Stakeholders are interconnected. Decisions made in one part of the system affect other parts, sometimes in unexpected ways. Sustainability challenges, environmental, social, and economic, require organisations to think beyond short-term efficiency. Sustainability is not an external addition to quality management but a natural extension of stakeholder-oriented thinking. When organisations systematically consider the expectations of all interested parties, environmental and social responsibilities become integral to the logic of quality. Systems thinking is a practical necessity and not an abstract academic concept.

Historically, this field has focused on defect reduction and operational efficiency. Later, customer satisfaction and process capability became the central focus. Today, the emphasis increasingly includes resilience, adaptability, and sustainability. Organisations

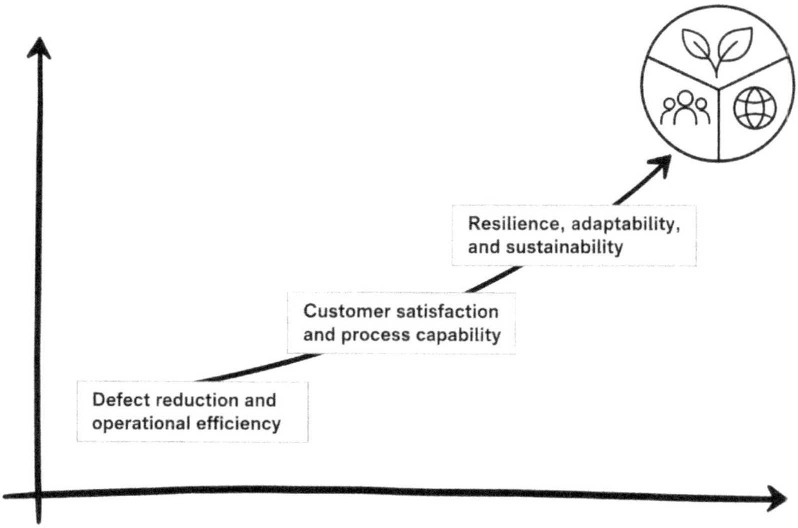

Figure 9.22 The evolution of quality management.

must be able to produce consistent results and respond intelligently to uncertainty and change.

Purpose becomes fundamental. Why does the organisation exist? Whose needs does it serve? What responsibilities does it have within its ecosystem? When the purpose is clear, improvement efforts gain direction. When the purpose is unclear, improvement risks becoming fragmented optimisation.

The future of quality management is likely to be shaped by several interrelated developments (Fundin et al., 2020)[3]

- Digitalisation and real-time data provide new possibilities. The processes can be monitored continuously. Trends can be detected early on. Predictive analytics can support proactive decision-making. Artificial intelligence can assist in identifying patterns that would otherwise remain hidden. However, technology does not replace judgement. Capability lies in transforming relevant information into responsible and timely actions.
- Expectations regarding sustainability will continue to increase. Organisations are increasingly evaluated not only on their financial performance but also on their environmental impact, social responsibility, and governance practices. Integrated management approaches and maturity models support broader accountability. Quality management can provide the structure and discipline required to navigate these expectations responsibly.
- Simultaneously, the world is becoming increasingly projectified. Many strategic initiatives, digital transformations, and sustainability efforts are implemented through projects. Improvement is often organised as structured, time-bound efforts embedded within stable operations. The project perspective discussed throughout this book is, therefore, a part of the practical expression of quality management in modern organisations.
- In such a world, stability and change must be coexistent. Control and improvement must reinforce each other. Quality management provides a framework that connects daily operations with purposeful development.

Ultimately, moving from compliance to capability means recognising that quality management is about strengthening the organisation's ability to act with clarity, coherence and responsibility. It also involves aligning the mission, vision, values, and culture, and ensuring that processes serve their purpose. From this broader perspective, quality management contributes not only to operational performance but also to organisational legitimacy and long-term societal trust.

The journey does not end with the certification. This does not end with a documented system. It does not end with project completion. It continues as long as the organisation chooses to learn. To become better and better is about commitment to purpose, capability, and responsible action in an evolving world.

This commitment defines the enduring relevance of quality management.

Case: Beyond Compliance in a Profitable Financial Consultancy

The company in this case is an international financial consulting firm specialising in investment advisory services for high-net-worth individuals and family offices. Its core business is to design tailored investment strategies, manage portfolios and provide strategic financial planning. The organisation operates in a highly regulated environment and handles extremely sensitive financial and personal data.

From an operational perspective, the company is well-managed. It consistently delivers strong financial results and distributes substantial dividends to its shareholders. Employees are well-paid, and recruitment is selective. Internal processes are disciplined, and risk management procedures are formalised. Owing to the nature of its operations, the company has implemented and maintained certification according to ISO 27001 for information security management. Data protection, confidentiality, and cybersecurity are critical priorities.

The owners take pride in their strict regulatory compliance. The firm operates within the boundaries of financial regulation and deliberately avoids legal "grey zones." All required reports were generated on time. The internal controls were documented. Audits did not reveal any major nonconformities. However, the external environment has changed.

Public debate in national and international media has increasingly focused on the widening gap between the rich and the poor. Financial advisory firms that serve wealthy clients are sometimes portrayed as contributors to structural inequality. Although the company has not been accused of wrongdoing, the tone of public discussion has become increasingly critical.

This debate has begun to affect employees' behaviour. Younger consultants, in particular, expressed discomfort when friends and family challenged the social value of their work. Some have raised internal concerns about the broader societal impact of a firm's activities, while others worry that reputational risks could affect long-term business stability.

Simultaneously, technological developments are reshaping the industry. The rapid development of AI and algorithm-based investment platforms has introduced new competitive pressure. Some analytical tasks traditionally performed by consultants can now be automated. Management has begun investing in AI-supported decision tools, but this has also increased employees' insecurity about future roles and competence requirements. Global financial markets have become increasingly volatile. Geopolitical tensions, rapid policy shifts, and technological disruptions have increased unpredictability. The firm's business model depends on trust, long-term relationships, and stable performance; however, the environment appears increasingly turbulent.

Discussions within the board of directors have intensified. Some owners argue that the company is already performing well; it is profitable, compliant, and efficient. Others suggest that this may no longer be sufficient. They refer to concepts such as social responsibility and legitimacy. Should the company articulate a clearer societal purpose? Should it expand its management system to address broader stakeholder expectations?

One proposal is to seek ISO 9001 certification. Given the existing structure and discipline of operations, certification could likely be achieved with limited effort. This would signal a formal commitment to quality management. Another suggestion is to explore broader frameworks of sustainability or responsible investment practices. The third perspective is that strategic reflection, cultural dialogue, and internal capability development are more important than additional certificates.

Therefore, the central question facing the company is not whether it complies with regulations. The question is whether compliance and profitability are sufficient foundations for sustained success in an ever-changing and more critical world. Is the next step the integration of broader management standards? A clearer articulation of purpose? Or is it something else entirely? Owners and managers must decide how the organisation should respond—not only to market volatility and technological disruption but also to evolving expectations about responsibility and value creation.

Reflection points—Chapter 9

1. Quality beyond the customer
   Quality has traditionally been defined in relation to customer requirements; however, future-oriented perspectives emphasise value for multiple stakeholders and society at large. Should organisations redefine quality to include environmental and social impacts, even if customers do not explicitly demand it? What tensions arise between market-driven quality definitions and broader societal expectations? Reflect on a manufacturing or financial organisation you know; does its quality approach extend beyond immediate customer satisfaction?
2. Purpose or performance optimisation
   If quality management becomes primarily a tool for efficiency and cost reduction, it risks losing its strategic direction and meaning. Should quality be anchored explicitly in organisational purpose and long-term contributions rather than short-term optimisation? What trade-offs emerge between measurable performance gains and less tangible dimensions, such as trust, legitimacy, and meaning? Consider an organisation in the public or education sector. How clearly is quality linked to its higher purpose?
3. Closed system or ecosystem thinking
   Many management systems focus on internal processes and boundaries; however, value creation increasingly occurs in interconnected networks. Should quality management expand to address system-level interactions across supply networks, partnerships, and communities? What governance challenges arise when responsibilities extend beyond formal organisational borders? Reflect on a company you know well, how does it manage quality across its broader ecosystem?
4. Efficiency or long-term sustainability
   Lean and continuous improvement emphasise waste reduction and efficiency, whereas sustainability introduces long-term environmental and social considerations. Can an organisation optimise efficiency without compromising future resilience and responsibility? How should leadership balance immediate productivity improvements with investments in sustainable capability? Apply this dilemma to an organisation in logistics or manufacturing with which you are familiar.
5. Predictive intelligence or human judgement
   The shift towards digitalisation and predictive analytics allows organisations to anticipate variations and detect anomalies in real time. Should decision-making increasingly rely on algorithmic insight, or must human judgement remain central to interpreting and acting upon data? What risks arise if predictive systems are implemented without corresponding ethical reflection and competence development? Reflecting on a healthcare organisation, how does it balance digital intelligence with professional responsibility?

6. Compliance culture or quality culture
   Certification and external audits can strengthen discipline, but they may also encourage a narrow-minded approach to compliance. How can organisations ensure that management standards support a living culture of improvement rather than procedural formalism? What trade-offs arise between demonstrating conformity and nurturing intrinsic commitment to quality? Consider a regulated organisation; does its management system primarily serve auditors, or does it genuinely strengthen its capability to become better?

Notes

1 Kume, H. (1987). *Statistical Methods for Quality Improvement*. Tokyo: Productivity Press.
2 Liker, J., & Meier, D. (2006). *The Toyota Way Fieldbook—A Practical Guide for Implementing Toyota's 4Ps*. New York: McGraw-Hill.
3 Fundin, A., Lilja, J., Lagrosen, Y., & Bergquist, B. (2020). Quality 2030: Quality management for the future. *Total Quality Management & Business Excellence*, 1–17. https://doi.org/10.1080/14783363.2020.1863778.

Index

For Product Safety Concerns and Information please contact our EU
representative GPSR@taylorandfrancis.com
Taylor & Francis Verlag GmbH, Kaufingerstraße 24, 80331 München, Germany

www.ingramcontent.com/pod-product-compliance
Lightning Source LLC
LaVergne TN
LVHW060317241026
850939LV00029B/419
* 9 7 8 1 0 4 1 1 3 8 3 3 4 *